MAD ON RADIUM

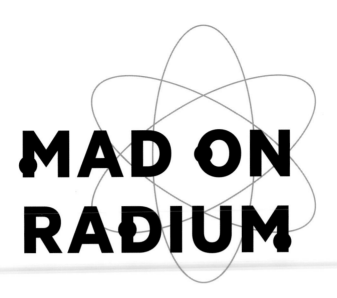

MAD ON RADIUM

NEW ZEALAND IN THE ATOMIC AGE

★

REBECCA PRIESTLEY

AUCKLAND
UNIVERSITY
PRESS

For Jonathan

First published 2012

Auckland University Press
University of Auckland
Private Bag 92019
Auckland 1142
New Zealand
www.press.auckland.ac.nz

ISBN 978 1 86940 727 8

Publication is kindly assisted by

National Library of New Zealand Cataloguing-in-Publication Data
Priestley, Rebecca.
Mad on radium : New Zealand in the atomic age / Rebecca Priestley.
Includes bibliographical references and index.
ISBN 978-1-86940-727-8
1. Nuclear energy—Government policy—New Zealand.
2. Nuclear energy—New Zealand—History. 3. Anti nuclear
movement—New Zealand. 4. Nuclear engineering—New Zealand—
History. I. Title.
333.79240993—dc 23

Front cover photograph: Neil Anderson,
Royal New Zealand Navy Museum, AAO0029
Cover design: Philip Kelly

Printed in China by 1010 Printing International Ltd

CONTENTS

★

Nuclear-free
New Zealand
Myth or reality?

. . . our nuclear free status . . . has become a defining symbol of our national identity. — PRIME MINISTER HELEN CLARK, 28 NOVEMBER 2007[1]

New Zealand's nuclear-free status is a myth . . .
— ACT MP KEN SHIRLEY, 27 JULY 2005[2]

Like most New Zealanders of my generation, I am a proud supporter of our country's stand against nuclear weapons. As a teenager in the 1980s I marched against visits by American nuclear warships to Wellington. When I travelled through the United States in the late 1980s, I was happy with the positive recognition that New Zealand's nuclear-free policy got me — and was just as happy to argue with people who took umbrage at our stand. But then my career exposed me to a 'nuclear' side of New Zealand that most of my contemporaries did not see. A few years after completing a degree in earth sciences, I worked for several months as a journalist for the Institute of Geological and Nuclear Sciences (IGNS), a Crown Research Institute that incorporated the former Department of Scientific and Industrial Research's (DSIR) Institute of Nuclear Sciences. For the rest of the 1990s, I continued doing contract work for IGNS, writing for annual reports, newsletters and marketing brochures, mostly about geological projects but sometimes about nuclear science. It was during this time that my friend Steve Menzies lent me his copy of Catherine Caulfield's 1989 book *Multiple Exposures: Chronicles of the Radiation Age*, and suggested there might be some New Zealand stories to uncover. I soon bought my own copy of the book and a seed was sown. It was from this position, and with a growing awareness that New Zealand *had* a nuclear history, that my interest in writing this story emerged. In 2001, I enrolled as a part-time PhD student in the University of Canterbury's history and philosophy of science programme.

The whole time I was enrolled, the most common response to telling people I was doing a PhD thesis on New Zealand's nuclear history was surprised laughter and the comment that 'it must be a very short thesis'. I was happy to prove them wrong — it would have been a disappointing thesis if I hadn't. As I trawled through the pages of government archives, read through old newspapers on microform readers, and interviewed retired scientists now in their seventies, eighties and nineties, the story took shape.

I was already familiar with New Zealand's nuclear-free story, one that has been told in many books, articles and documentaries. New Zealand's nuclear-free identity is strong, and most New Zealanders support the nuclear-free policy. From a grassroots movement that began with opposition to nuclear bomb testing in the Pacific and later extended to opposition to visits from nuclear-armed and nuclear-powered warships and the creation of local 'nuclear-free zones', the New Zealand Nuclear Free Zone, Disarmament and Arms Control Act of 1987 created a nuclear-free zone that prohibited nuclear weapons and nuclear warships in the country's land, air and water. In the years that followed, governments and the New Zealand public began to interpret the nuclear-free policy more widely. In 1996, a Minerals Programme issued under the Crown Minerals Act prohibited uranium mining and prospecting for uranium. And in 2005, in response to calls for nuclear power to meet New Zealand's future electricity needs, the Labour Government stated their clear policy of no nuclear power stations. This nuclear-free status, initiated by a Labour Government in the 1980s, is now recognised by all major political parties, and is a reflection of a strong nuclear-free national identity.

But I discovered over the course of my research that, while being nuclear-free is now hugely important to most New Zealanders, it is different from other posited national traits (like a talent for ingenuity or a passion for rugby) in that it is a relatively recent addition to our national story. My research revealed that — apart from New Zealand's public and governmental opposition to nuclear bomb testing in the Pacific — a broad all-encompassing 'nuclear-free' ethos was not strongly apparent before the events of 1985, when Prime Minister David Lange refused to allow the USS *Buchanan* access to New Zealand ports, and when French agents bombed the Greenpeace ship *Rainbow Warrior* while it was berthed in Auckland. I found a lot of early enthusiasm for nuclear science and technology: from the first users of x-rays and radium in medicine; the young New Zealand physicists seconded to work on the Manhattan Project; the plans for a heavy-water plant at Wairakei; prospecting for uranium on the West Coast of the South Island; plans for a nuclear power station on the Kaipara Harbour; and the thousands of scientists and medical professionals working with nuclear technology. Put together, they provided an alternative history of a nuclear New Zealand.

By examining New Zealanders' consideration of aspects of nuclear science and technology — like nuclear medicine, nuclear power, and uranium prospecting — and examining the reasons why they were or were not adopted or accepted, I discovered that throughout most of the twentieth century it was economic considerations that tended to dominate public and official attitudes and decisions when it came to things nuclear. There is nothing unusual about this: Australia and Canada developed uranium-mining industries because it was economically advantageous; New Zealand did not develop a uranium mining industry because it was economically unviable. But this led to the realisation that, had economic factors suggested different decisions be made, New Zealand might not today have such a strong nuclear-free identity. In the late 1970s, less than a decade before becoming so proudly nuclear-free, New Zealand was considering nuclear power to meet the growing electricity demand in the North Island. In 1978, a Royal Commission of Inquiry rejected the immediate development of nuclear power for New Zealand, not in response to anti-nuclear sentiment, which did exist, but because a reduction in projected electricity demand and the recent discovery of the Maui natural gas field meant New Zealand had sufficient indigenous resources to meet electricity needs until the end of the century. I also discovered a previously untold story of the history of uranium prospecting in New Zealand, and found that, while New Zealand never mined uranium, this was not because of a moral decision not to provide materials for the international nuclear power and weapons industries. Rather, it was because no economically viable deposits of uranium were ever found, despite 35 years of uranium prospecting initiated by the DSIR and supported by the New Zealand and British governments.

I believe that a lack of acknowledgement of our nuclear history, along with an unwillingness to discuss nuclear issues, is one reason there is such suspicion of anything with nuclear associations in this country. While New Zealand's nuclear-free policy now refers to nuclear weapons, nuclear propulsion, and nuclear power, other practices associated with radiation or nuclear science have been tainted with the 'nuclear' association. In 1995, Hutt City mayor Glen Evans proposed that the Hutt City Council declare its city a nuclear-free zone as had the former Lower Hutt City Council in 1984. A New Zealand Press Association (NZPA) report declared that Hutt City could not do this 'because of the nuclear reactor at the Gracefield nuclear physics research centre'.[3] Evans and the NZPA were not the only people confused. It was a common misconception that there was a nuclear reactor at the DSIR's Institute of Nuclear Sciences, by now part of the Institute of Geological and Nuclear Sciences. In the early 1990s when I was working for the institute (now known by the less provocative moniker of GNS Science), taxi drivers would call it 'the bomb factory' and warn me about the nuclear reactor on the Gracefield hill site (actually a particle accelerator used for radiocarbon dating). When Jools Topp, of the Topp Twins country music duo, was offered radiation therapy as part of treatment for breast cancer, she refused it, telling Radio New Zealand's Kathryn Ryan in 2009 that she had advised the doctors 'I'm a lifelong member of Greenpeace, why would I let you irradiate me?'[4] But New Zealand society relies on nuclear technology much more than most people realise, with more than $4 million spent each year importing radioactive isotopes for use in industrial and environmental applications as well as in medicine.

There have been challenges to New Zealand's nuclear-free policy in recent years, and there will be more challenges in the future. Human-induced climate change and the need for new sources of energy with a low carbon footprint have led to calls for nuclear power to be revisited as an option for New Zealand. Both former National Party leader Don Brash and former prime minister Geoffrey Palmer have suggested in recent years that New Zealand revisit the question of allowing nuclear-powered ships into ports. But many of the voters who might be considering these options in the future are from younger generations unaware that we ever considered the nuclear power

After public and media response to Don Brash's comments that under a National Government New Zealand's nuclear ships ban would be 'gone by lunchtime', he quickly tried to backtrack on this policy. Cartoon by Tom Scott, *The Dominion Post*, 24 Jun. 2004. A-312-4-023, Alexander Turnbull Library, Wellington, New Zealand.

In April 2010, former Labour Prime Minister Geoffrey Palmer suggested that New Zealand might revisit the issue of visits from US nuclear warships. In 1984 Palmer had been deputy prime minister of the Cabinet that advised David Lange against accepting a visit from an American warship that would not conclusively confirm or deny whether or not it was nuclear-armed. Cartoon by Malcolm Evans, *The New Zealand Herald*, 21 April 2010, www.evanscartoons.com.

option for New Zealand in the past, or who do not know why we banned nuclear-armed and nuclear-powered ships from our ports.

While New Zealand's nuclear history has been overshadowed by the well-told 'nuclear-free' story, I am not the first person to tackle this subject. During the ten years I was researching and writing about New Zealand's nuclear history, two informative books on the subject were published: retired health physicist Andrew McEwan launched the argument that New Zealand is not 'nuclear free' in his 2004 book *Nuclear New Zealand*, which covered some aspects of New Zealand's nuclear history but mostly served to meet McEwan's goal of clearing up what he described as public misconceptions about nuclear and radiation issues in New Zealand; and retired diplomat Malcolm Templeton looked at New Zealand's nuclear history from 1945 to 1990 as a foreign policy issue in his 2006 book *Standing Upright Here*. My work draws on earlier publications by Ross Galbreath, whose book *DSIR: Making Science Work for New Zealand* includes a chapter on the DSIR's Institute of Nuclear Sciences; by defence historian John Crawford, who has written about New Zealand's role in the British nuclear-testing programme; and on Andrew McEwan's 1983 history of the National Radiation Laboratory.

I worked on the thesis on which this book is based for the best part of a decade, and many people and organisations assisted me along the way. First thanks go to the University of Canterbury, whose doctoral scholarship provided vital funding over the first few years of my research, and to my supervisors at the University of Canterbury: Philip Catton (philosophy), Professor John Hearnshaw (physics and astronomy) and Professor Philippa Mein Smith (history). Second thanks go to the Stout Research Centre for New Zealand Studies at Victoria University of Wellington, and its director, Professor Lydia Wevers, for giving me a peaceful haven in which to write. Thanks also to my external PhD examiners, Associate Professor Ruth Barton from the University of Auckland, and Professor David Miller from the University of New South Wales, whose comments on my thesis helped to guide me as I adapted it into a book. Further thanks go to the Ministry of Culture and Heritage for providing me with a History Research Trust Fund Award to turn the finished thesis into a manuscript for publication.

Many individuals assisted me by providing personal anecdotes, by sharing photographs, diaries and papers or by sharing their expertise on specialised topics. Particular thanks go to Jim McCahon, Robin Williams, Jack Tait,

Lloyd Jones, Rob Aspden, Paul Cotton, Tony Silke, Derek Olsen, Jeremy Whitlock, Jock Brathwaite, Simon Nathan, the family of Tas McKee, GNS Science, Greenpeace, the National Radiation Laboratory, Gregory Good from the American Institute of Physics, Jonathan Ward from the National Research Council Canada, Tom Scott, Malcolm Evans, Rodger Sparks, Piers Anderson, Chris Lonsdale and the Lodge family. And thanks to the staff at Archives New Zealand and the National Library of New Zealand, where I spent many weeks, probably months, researching. Every effort has been made to find and clear permissions with all copyright holders; if any oversights have been made I would welcome hearing from rights holders.

Special thanks go to the wonderful team at Auckland University Press — especially Sam Elworthy, Anna Hodge, Katrina Duncan, and Christine O'Brien — and their freelancers, particularly Kate Stone, Philip Kelly and Louise Belcher. And final thanks go, of course, to my family.

Rebecca Priestley
August 2012

1

*

The public are
mad on radium!
Rutherford, New Zealand
and the new physics

. . . The public are mad on radium . . .
— GOVERNMENT BALNEOLOGIST ARTHUR WOHLMANN, 1914[1]

*The energy produced by the breaking down of the atom is a very poor
kind of thing. Anyone who expects a source of power from the
transformation of these atoms is talking moonshine.*
— ERNEST RUTHERFORD, 1933[2]

The Rotorua Bathhouse, a large and elaborate complex in the style of
a European spa, was for many years the focus of the small town that
grew around it. From when it opened in 1908, weary or chronically ill
ladies and gentlemen would come to the bathhouse to soothe their arthritic
joints with massage and thermal baths, or to smooth mud on their eczema- or
psoriasis-ravaged skin. After 1914, they could also treat their depression or
constipation with a course of specially prepared glasses of radioactive water.

The enterprising idea of adding radon water to the spa's list of treatments
came from Arthur Wohlmann, the balneologist — an expert in the therapeutic
use of mineral waters — responsible for New Zealand's government-owned
spas at Rotorua, Te Aroha and Hanmer. After a trip to Europe, Wohlmann

In 1914, the luxurious Rotorua Bathhouse added radioactive water — four to six glasses a day were recommended — to the spa's list of therapeutic treatments. OP-2489, Rotorua Museum of Art and History, Te Whare Taonga o Te Arawa, Rotorua, New Zealand.

had declared that treatment by radioactive waters, with their 'alterative effects on metabolism', had come to stay and its 'possibilities were very great'.[3] Wohlmann described radon water as being able to be provided by injection, by vaginal or rectal douches, or insertion into tooth cavities, but at the Rotorua Bathhouse he focused on administering radon through the skin and lungs, by soaking in steaming hot radioactive baths or inhaling radon gas, and by drinking radioactive water, which he described as 'by far the most satisfactory method of administration' as it stayed in the body much longer.[4]

For £250 — quite a sum in 1914 — he bought a 'radium activator': a porcelain jug with a side tap for draining off water. Inside the jug, a small container held a minute quantity of the mineral salt radium bromide, from which a continuous emanation of radon gas irradiated any water used to fill the jug. The radioactive water was drawn off and replenished daily. The radium, with a half-life of 1620 years, could irradiate the water almost indefinitely. But the radon gas — the element that gave the water its radioactive 'curative' properties — had a half-life of less than four days, and so for the treatment to be 'effective' patients were advised that the radioactive water had to be drunk within 24 hours of being drawn. Wohlmann recommended that each patient take four to six small glasses a day, with the radioactive water promoted as being especially valuable in treating gout, diabetes and constipation, as well as for soothing the nerves and, according to the local newspaper, 'tightening loose teeth'.[5] As word of the new therapy spread, many patients were sent to Rotorua for radon treatment: in 1916, more than 8500 glasses of radon water were sold.

To imbibe radioactive water as a 'treatment' for chronic illness seems preposterous in light of what we now understand about it, but Wohlmann was no charlatan: he was a medical doctor and a scientist, striving to keep up to date with, or ahead of, the rest of the world when it came to the latest applications of his science. In giving radioactive water to his patients, Wohlmann was acting on scientific beliefs based on work by Marie Curie, and others, that suggested that moderate doses of radioactivity promoted 'the multiplication and growth of healthy cells and the decay of morbid ones'.[6] His patients, in turn, were seeking the assurance of having the latest treatments on offer. New Zealanders in the early twentieth century were, therefore, as enthusiastic about subjecting themselves to unnecessary radiation as people from any Western nation. While some harmful effects of radiation were already evident, they were not well understood, and radiation technologies had proved so beneficial in the diagnosis of disease and the treatment of cancer that radon water treatment was, for many, seen as a natural extension of the benefits of this marvellous yet poorly understood new phenomenon called radiation.

A NEW SCIENCE IS BORN

When Wohlmann began offering radon water treatment, in 1914, it was less than two decades into the revolution in physics that had begun with the discovery of first x-rays, and then naturally occurring radioactive elements like radium. It began in 1895, when German physicist Wilhelm Röntgen observed the mysterious and invisible electrically generated rays, which he called 'x-rays', that could pass through paper, wood, rubber, copper and even thin sheets of most metals, but not through bones or lead. By placing opaque objects between the source of the rays and a photographic plate, Röntgen discovered he was able to take x-ray pictures.

News of Röntgen's discovery spread quickly to the popular press, where it caused an international sensation. Responding to public excitement about the discovery, Röntgen appeared at packed public lectures to demonstrate the workings of his x-ray machine, training it on the skulls, arms, and legs of enthusiastic volunteers. Some people were alarmed by his discovery. There was talk of banning x-rays in opera glasses for fear of insulting the virtue of the female singers by seeing through their clothes — one company even started to market x-ray-proof underwear. Others were quick to take up the relatively simple science involved. Doctors saw the potential to use x-rays as a diagnostic tool, and in February 1896 Canadian surgeons used x-rays to help locate and remove a bullet from a man's leg. Doctors also began to use x-rays for ostensibly therapeutic purposes — to treat dermatitis, cancer and tuberculosis.

In New Zealand, the media — concerned mainly with colonial news, the price of mutton in London, and the latest shipment of British goods to arrive at Ballantynes or Kirkcaldie & Stains — were more subdued, responding only after the discovery of x-rays had made news in Britain. The *New Zealand Mail* reported that London doctors were using Röntgen's discovery to take pictures of gallstones and injuries to the bones, achieving 'astonishing results'.[7] The new discovery caused great excitement in the New Zealand medical and scientific communities, and x-rays were discussed at meetings of the New Zealand Medical Association and at the scientific societies that made up the New Zealand Institute. *The New Zealand Medical Journal*'s London correspondent described the new photography developed by Röntgen for New Zealand readers, exclaiming that never 'has a scientific discovery excited more general interest, been followed up with such rapidity, and attained such extended success'.[8] Lawyer and naturalist William Travers, in his 1896 presidential address to the Wellington Philosophical Society, provided a detailed account of the discovery of 'Röntgen Rays', which he described as 'a most remarkable event in the history of physical and chemical science'.[9] In the years that followed, first x-rays, and then radium and radioactivity, were occasional topics for popular lectures, always with experiments and demonstrations, at universities and scientific and philosophical societies around New Zealand.

But the New Zealand medical profession did more than just talk about the new discovery, they started to use it. The technology was easy to reproduce: all that was needed was 'one spark coil, one battery, one Crookes tube and some

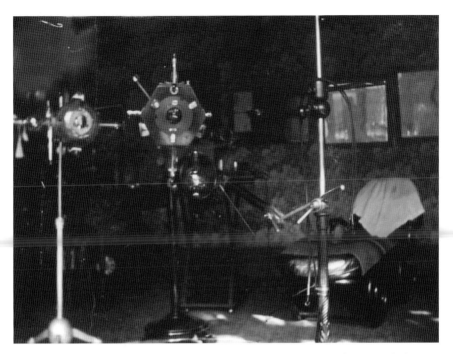

William Hosking, medical superintendent of Masterton Hospital, was one of the first New Zealand doctors to offer x-ray diagnosis and therapy. He offered his services, and later radium therapy, to paying customers from a clinic set up in his home. 02-6/7.digital, Wairarapa Archive, Masterton, New Zealand.

facility in handling photographic plates',[10] and some medical practitioners and businessmen were quick to experiment with the new technology and start charging for its use. Just months after Röntgen's discovery made the papers, New Zealand's largest drug and fertiliser company, Kempthorne, Prosser and Co., began establishing a Dunedin laboratory offering x-ray photography to the medical profession. In August 1896, they demonstrated the new technology to Dunedin doctors at a local meeting of the New Zealand Medical Association. In the North Island, William Hosking, medical superintendent at Masterton Hospital, imported a six-inch coil that he installed at his home, and began offering diagnostic and therapeutic x-rays to paying customers. For Hosking, and many of these other early adopters, the new technology pre-dated the supply of mains power, and an outdoors generator or battery was needed to power the electrical equipment.

Other entrepreneurs and doctors followed, finding the new technology particularly useful for identifying broken bones and lung disease. Hospitals initially contracted radiological services from local doctors or electricians who had purchased and installed their own x-ray equipment, but by 1898 the Auckland, Wellington and Christchurch hospitals each had their own x-ray equipment. Dunedin Hospital followed in 1904. It wasn't just doctors who found x-rays fascinating and useful. Amateurs and hobbyists were also attracted to the new technology. One North Otago sheep farmer found the battery-powered six-inch spark coil he imported in the late 1890s useful for 'radiographing dogs' legs, locating foreign bodies, and even the examination of broken wrists'.[11]

But these early x-ray machines, while incredibly useful — their ability to see inside the human body was extraordinary for the time — were also very dangerous. The primitive equipment, poorly understood physics, and a lack of comprehension of the medical effects of x-rays meant that radiation burns and even electric shocks were commonplace. While a simple chest x-ray today usually involves radiation exposure of less than one second and provides a radiation dose equivalent to ten days' natural background radiation, early x-rays came from much weaker and more primitive equipment: they took a long time and, while they were inconsistent in the amount of radiation they produced, they delivered much higher radiation doses than today. Pioneering Dunedin radiologist Harry de Lautour — who between January 1899 and October 1900 took 157 'radiographs' of patients — advised x-ray exposure times of 'four or five minutes for a hand or foot; eight or ten minutes for an ankle, leg or forearm; twenty to twenty-five minutes for a thigh, shoulder or chest', reassuring the reader by saying: 'so far I have not yet had any experience of burning'.[12] How the patients managed to keep still for that long, he doesn't say.

DISCOVERY OF RADIOACTIVITY AND RADIUM

Röntgen's discovery of x-rays had set off a chain reaction in physics research. First, French physicist Henri Becquerel discovered that uranium — a mineral used to colour pottery and glass a range of yellows and oranges — was spontaneously emitting rays similar to Röntgen's x-rays. Physics student Marie Curie then began testing other elements, and found that thorium, as well as uranium, emitted the rays, which she called 'radio-activity'. Working at the Cavendish Laboratory at the University of

Cambridge, New Zealander Ernest Rutherford concluded, after studying the radiations emitted by uranium and thorium, that uranium emitted 'at least two distinct types of radiation — one that is very readily absorbed [it could be stopped by a piece of paper or a few centimetres of air], which will be termed for convenience the α [alpha] radiation, and the other of a more penetrative character, which will be termed the β [beta] radiation'.[13] In 1900, French physicist Paul Villard discovered a third type of radiation, a form of high-energy penetrating x-rays, which he named gamma (γ) radiation. Rutherford later demonstrated that beta radiation was a stream of negatively charged electrons and that alpha radiation consisted of positively charged helium atoms ejected during radioactive decay.

Curie continued her research into radioactive elements, and, after years of painstaking work, she and her physicist husband Pierre found that pitchblende, a black, shiny ore from which uranium was extracted, contained two previously unknown elements, polonium and radium, each of which was more radioactive than uranium. The remarkable thing about radium was its incredible energy output — each gram of the highly radioactive element could inexplicably heat a gram of water from freezing to boiling point in less than an hour.

By now, Rutherford was working in Montreal where, with chemist Frederick Soddy, he discovered that in the process of emitting radiation an element is spontaneously transformed into another element. This remarkable discovery — a transmutation which they described as 'modern alchemy' — helped to explain the seemingly inexhaustible supply of energy from radioactive elements like radium. Rutherford and Soddy also discovered that all of the radioactive elements had a distinct 'half-life' — the time it takes for half of the atoms of the original sample of an element to decay into a new element. The half-lives of the elements they tested varied wildly: uranium's half-life was calculated at 4.5 billion years, radium's half-life was 1620 years, and a decay product of thorium had a half-life of only 22 minutes. Other decay products were found to have half-lives of only fractions of a second.

Compared to x-ray technology, radium was, for most scientists and doctors, prohibitively expensive. Even so, several New Zealanders managed to get samples of radium soon after it began being manufactured in Europe, mostly by virtue of connection with Rutherford or other European scientists.

WIDESPREAD USE OF X-RAYS IN NEW ZEALAND

Over the following decades, an enthusiastic medical profession and paying public ensured that first x-rays and then radium became an integral part of mainstream medical diagnosis and treatment. Awareness and understanding of the hazards of working with radiation, however, were slower to develop.

The new technologies of x-rays and radium were linked, in that they both produced penetrating (and what we now know as ionising) forms of radiation that could be used in diagnostic or therapeutic medicine. While the public were unlikely to be aware of the physical differences between, for example, x-ray therapy and radium therapy, they were in fact two very different processes: an x-ray machine produced an electrically generated form of electromagnetic radiation emitted by electrons that could be used in diagnosis or therapy; radium, or its daughter product, radon gas, was used mostly as a close-range therapy for the alpha particles it emitted. Radium had applications outside of medicine, too: when mixed with beryllium, scientists could use it as a source of neutrons for physics experiments.

By the second decade of the twentieth century, diagnostic x-rays were in widespread use in New Zealand's hospitals and dentists' rooms. However, specialised medical attention like x-rays came at a cost to the patient. An x-ray 'radiograph' cost between half a guinea and three guineas in 1917 (up to $100 in 2012 New Zealand dollars), thus limiting its application to wealthier patients. The First World War saw the development of new x-ray apparatus and the standardisation of x-ray techniques, most of which reached New Zealand by 1920. The technology was becoming more sophisticated, but New Zealand, with its small population, did not yet have the trained professionals to use it. While some hospitals employed lay radiographers, x-ray equipment was just as likely to be used by hospital engineers, electricians and nurses — whoever was available at the time. One nurse described taking her knitting into the x-ray room: 'it worked out something like half a row of sock for an ankle and up to two rows for a lumbar spine'.[14]

As the technology became more advanced and established, medical applications for x-rays expanded beyond diagnosing bone fractures and joint conditions. Chest x-rays became an important tool for confirming the diagnosis and extent of pulmonary tuberculosis, which only a few decades earlier had been the number one killer of New Zealanders. By injecting or feeding patients with solutions that were impervious to x-rays, such as

WE HAVE PLEASURE IN BRINGING UNDER YOUR NOTICE THE

Vulcan Portable X-ray Equipment.—£100.

Perfectly Simple—Simply Perfect.   Every Professional Man his own Operator.

This equipment is sold on a positive guarantee to produce the highest class of Chest Radiograph in 1½ seconds', Hip Radiograph in 2 seconds' exposure, &c., as fine as can be produced with any apparatus regardless of size or price.

NO SUBJECT BEYOND ITS RANGE.   Sample Radiographs and full information on request.

Sole Distributors for New Zealand: **J. and W. FAULKNER,** Castle Street, DUNEDIN. P.O. Box 565.

From a handful of x-ray machines at the turn of the century, New Zealand had an estimated 450 x-ray installations by 1944. This advertisement ran in the *New Zealand Journal of Health and Hospitals* in 1920. H1, box 1816, 53/119, alt 28289, Archives New Zealand, Wellington, New Zealand.

compounds containing barium or iodine, soft tissues such as the digestive and urinary tracts could be examined by x-ray.

Today, the developing fetus is understood to be vulnerable to any form of ionising radiation, and exposure of pregnant women to x-rays and other forms of avoidable radiation is minimised. But in the 1920s, decades before ultrasound scanning technology was developed, x-rays offered an exciting and unprecedented way to observe the fetus. In 1926, by which time 44 New Zealand hospitals had x-ray equipment, the Director-General of Health proudly announced that arrangements had been made for pregnant women to have x-rays for the diagnosis of conditions such as multiple pregnancy, hydrocephalus and malformation of the fetal skeleton. The practice of offering antenatal care to all pregnant women was new, launched only two years earlier as part of the Department of Health's Campaign for Safe Maternity. With the new x-ray screens and (for the time) fast films now in use, the Department of Health assured there was 'no danger either to mother or child'.[15] By the mid-1930s the antenatal x-ray was on the list of standard x-ray procedures offered by New Zealand hospitals, available for an outpatient charge of 5 shillings a film.

X-rays were popular and the public was happy to pay for them: as well as being offered in hospitals, x-ray machines were used in health spas (to diagnose joint conditions) and by chiropractors. In shoe shops, sales staff with no training in radiography operated 'pedascopes' or 'shoe-fitting fluoroscopes'. Most pedascopes had no limits on radiation exposure time, and children could play unsupervised on the machines, irradiating their feet and watching their foot bones on the screen.

X-RAY AND RADIUM THERAPY

X-rays were used diagnostically as soon as it was discovered that the rays did not pass through human bones, but their therapeutic use followed experiments to determine their effect on skin diseases. X-rays were perceived to have a beneficial effect on skin conditions like acne, ringworm and skin cancer, although it was soon noticed that radiation also caused skin burns and hair loss.

As with the adoption of diagnostic x-rays, experiments with radiation therapy in New Zealand followed close behind the first international publication of the new techniques. Radiotherapy trials in New Zealand began in 1901, but more advanced therapy was taking place in London, and some New Zealanders travelled there for treatment. In 1902, *The Press* described how a Christchurch patient, Craig Robertson, who had suffered for 26 years from 'rodent ulcer' (now known as basal cell carcinoma), was treated in London with the new 'X-Rays Light Cure'. Before leaving for London, Robertson had been through fifteen operations to treat his condition, which affected his face, and the 'ulcer, which started in his cheek, just to one side of the mouth, had extended close to his eye . . . and he was in great fear that it would go to the brain'.[16] On arriving in England, Robertson received treatment at London Hospital. *The Press* described his treatment:

> The light . . . is administered through a round globe . . . . The patient sits down before the battery, with the globe placed on a wooden stand at about the level of his head, and from twelve to eighteen inches in front of him. Two small coils

are attached to points at the two ends of the globe, from the battery. A leaden mask is placed over the patient's face, with a small hole cut in it just over the place where the ulcer is, and the apparatus is placed so that the light from the X-Rays will fall exactly upon that spot. The mask is used to prevent the rest of the face being shrivelled away. A tap is given to the battery in the manner of touching a spring, and the process begins. When it has lasted for ten minutes a second touch with the finger cuts off the light. An exposure of the ulcerated place to this light during ten minutes each day, for six days a week, forms the whole of the simple process.[17]

Following his treatment, which involved ten-minute x-ray sessions, six days a week, for several months, Robertson declared, 'I felt noticeably better after only eight days, and in from five to six months I was quite well.'[18]

Enthusiastic reports like Robertson's promoted x-ray therapy as a near-miraculous treatment, and New Zealand was keen to keep up to date with the latest therapies being offered in the United Kingdom. New Zealand medical practitioners were quick to master the new technology, and it was not long before individual practitioners in New Zealand were offering paying patients radiotherapy services. In Masterton, Hosking added radium and x-ray therapy to his private x-ray diagnostic services, and it was reported that he used radium to cure a carcinoma of the lip soon after the turn of the century. When it came to therapy, an advantage of radium over x-rays was that it could be used to treat cancers that were difficult to reach externally, and was therefore used for inter-cavity treatments, like cancer of the uterus, or for direct insertion into tumours. By emitting alpha particles, which travel only a few millimetres, the radium was able to destroy the cancerous tissue into which it was inserted without damaging the healthy tissue surrounding it. Radium, however, was much more expensive than x-ray equipment and was hard to obtain, and so it took longer to become established for medical use.

Marie and Pierre Curie had first extracted radium in 1902, and by 1907, in recognition of its medical benefits, radium was being extracted in one Austrian and two French factories. By 1913, however, they had together made available only 20 grams of radium. The United States were the next to enter the market, and between 1913 and 1926 they put about 200 grams of radium on the market, about half of which was used in medicine, with the rest being used in luminous paints. When Belgium and Canada entered the radium market in the 1920s, this precious substance became more available and affordable to hospitals in New Zealand, with the price dropping from £33,000 per gram in 1913 to £25,000 per gram in 1926, and then becoming progressively cheaper once radium mined from the Belgian Congo began to dominate the market.

Although radium was becoming increasingly available, New Zealand hospitals were poorly funded. Under the Hospitals and Charitable Institutions Act 1909, hospitals were funded by levies on local bodies, central government subsidies, and monies from donations, bequests and patient fees. Public appeals were, by necessity, a common way of raising money for capital expenditure. In 1914, Percy Cameron, who was in charge of Dunedin Hospital's x-ray department, recommended the establishment of a 'Radium Institute' at the hospital. Cameron had bought a small supply of radium in 1911 and was using it successfully at the hospital and in private practice. The same year, Herbert Inglis, honorary radiologist to Christchurch Hospital, made an appeal for public funds to buy a supply of radium for the hospital.

Medical opinion, however, was divided on the success of radium as a therapy, and in Wellington the hospital's medical superintendent cited the 'enormous cost of radium . . . The differences of opinion amongst eminent medical authorities as to its therapeutic value, the dangers attending its use, and the absolute necessity of having a trained expert for its administration' as reasons to hold over a proposal to secure a supply of radium for Wellington.[19]

By 1917, however, medical and public opinion had moved in favour of radium. Public appeals first in Dunedin, then in Wellington, Christchurch and, finally, Auckland, raised money for the purchase of radium supplies. With additional funding from the Government, radium therapy departments were established in New Zealand's four main hospitals. Wellington's radium supply, personally selected by Ernest Rutherford from Radium Belge's supplies on sale in London, arrived at Wellington Hospital, along with an authentication certificate signed by Marie Curie. Most of the hospital's radium supply — 650 milligrams — was installed in a radon plant in the hospital's basement. Radon gas emanating from the radium was collected and sealed in tiny glass tubes before being enclosed in suitable applicators, including platinum seeds and needles, for use in cancer treatment throughout the country.

In Christchurch, P. Clennell Fenwick, head of the hospital's x-ray department, travelled to the United Kingdom to train in radium therapy. Such was the demand for x-ray therapy and radium treatment that by early

CANCER

HELP FOR THOSE WHO SUFFER

An appeal for funds to provide a Radium Department at the Wellington Hospital for all sufferers with this dread disease, in the territory served by the Wellington Centre.

" How dead the soul to human duty which does not respond to a cause so noble."

" No greater joy will ever come to you than to know that some of the wealth which God has entrusted to your keeping has been used for the relief of your suffering fellows."

Wellington's radium appeal successfully appealed to citizens' sense of duty in raising £10,000 for the purchase of radium and radiotherapy equipment for Wellington Hospital. MS-Papers-1293-119/03, Misc Records, Hutt City Council, Alexander Turnbull Library, Wellington, New Zealand.

1925 Fenwick said he was treating patients all day 'and all night too' at Christchurch Hospital's radium department.[20] Just two years after the department was established, Fenwick was finding it very difficult to keep up with the demand for radium therapy; he described the hospital's radium supply as being in 'incessant use'.[21]

Radium, the subject of public appeals in the four main centres, had needed little introduction: as Rotorua Bathhouse balneologist Arthur Wohlmann said, the public were 'mad on radium'.[22] As well as being known for their medical use, x-rays and radium had become part of popular culture. A 1900 advertisement for a popular brand of peppermints proclaimed:

The unassuming Rontgen Ray
Appears to burn the flesh away
And leave the white and ghastly bones,
The cause for shudders, sighs, and groans;
So like a man who is ill and cold,
Who thinks he's dead until he's told
The way to health in manner sure
By taking Woods' Great Peppermint Cure.[23]

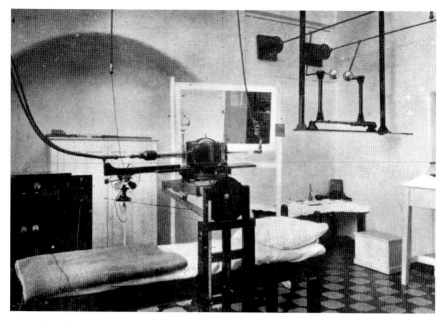

Christchurch Hospital's x-ray department, as shown in P. C. Fenwick's *History of the North Canterbury Hospital Board*. B-K 675-28, Alexander Turnbull Library, Wellington, New Zealand.

Other manufacturers used references to x-rays and radium to give their products an air of modernity and superior strength. From as early as 1911, Radium Polishes Ltd offered a range of polishes (none of which contained radium). As well as Radium Floor Polish with which 'everything will be brighter', New Zealand housewives could clean their stoves with X-ray Stove Polish with 'The Shine That Lasts Longest' and, somewhat alarmingly, bake bread using Radium Brand Flour.

Public lectures on x-rays and radium continued to attract crowds of curious non-scientists. In 1919, lectures on 'The Discovery and Properties of Radium' and 'The Lessons of Radium' by Professor Gwilym Owen, Auckland University College's physics professor, were packed, with the hall's 400 capacity not big enough to accommodate the enthusiastic crowds.

Once full radiotherapy services were established in the four main centres, the hospital boards worked to add to their supplies of radium and keep their equipment up to date. At first, William Massey's Reform Government matched public donations to the tune of 10 shillings in the pound. Lobbying from the hospital boards led the government of Gordon

Advertisers used the connection with radium and x-rays to confer superior cleaning power on their polishing products. *Auckland Star*, 25 Nov. 1908, p. 10; *The Dominion*, 1 Nov. 1923, p. 2.

Coates, who succeeded Massey, to match donations pound for pound to a total of £5,000 of government money for each of Auckland, Wellington, Christchurch and Dunedin hospitals. Patients outside these centres needing treatment could either have x-ray therapy from local equipment, or if 'deep therapy' was required they could travel to the nearest main centre or have an appropriate stock of radon sent out for their treatment. The maximum fee for a complete course of deep therapy was set at £25.

Through this combination of private donations and government support, New Zealand's desire to keep up with the rest of the world paid off. By 1929, the combined radium stocks of Auckland, Wellington, Christchurch and Dunedin hospitals (in the form of pure radium, radium needles and radium plaques) was such that New Zealand had a greater supply per head of population than the United Kingdom. Wellington Hospital, with 650 milligrams of radium in the emanation plant plus 100 milligrams of radium available for use in needles and applicators, had the greatest supply. Smaller amounts of radium were held by Wanganui Hospital, Victoria University College, and by independent medical practitioners around the country. Following consultation with radiologists using radium, the Department of Health decided that each main centre should have a 1-gram supply. By this time radium was selling for £12,000 per gram — far cheaper than its 1913 price of £33,000 per gram — but it was still, as Wellington's *Dominion* newspaper described it, 'more costly than the rarest jewels'.[24]

The mechanisms by which radiation worked as a cancer treatment — and therefore the potential dangers of its use — were as yet not well understood by the public, nor even by some medical professionals. If radiation was such a successful treatment for cancer, surely it could be used for other conditions? Enthusiasm for radium led both unscrupulous quacks and well-meaning medics to take advantage of the new therapy, prescribing first x-rays and then radium for all manner of ailments. Respected Dunedin Hospital radiologist Colin Anderson described radium being used in the 1920s to treat women with non-malignant conditions such as fibroids, or even irregular or excessive menstruation; the standard treatment was to insert a 50-milligram tube of radium bromide into the uterus for 24 hours. Anderson noted that, 'the haemorrhage in young girls is even more difficult to control than in older women' and suggested that 'fairly large doses may be given without fear of destroying ovarian function'.[25]

New Zealand, however, escaped some of the most bizarre applications of radium. While radiation was used by New Zealand dermatologists to treat benign skin conditions like acne, ringworm and birthmarks, internationally, radioactive products were also sold to 'treat' myopia and arthritis, and for cosmetic purposes such as banishing unwanted facial hair. Radium was also introduced to products as unlikely as toothpaste, contraceptive gels, face creams and even chocolate bars.

New Zealanders who read magazines and newspapers from abroad would have been aware that radon, the radioactive gas that emanated naturally from radium, was being promoted in Europe and the United States as a cure-all and general tonic, often in the form of radon water. At the Rotorua Bathhouse sales of radon water peaked in 1916, when more than 8500 glasses were sold. After that, there was a steady decline in sales, with returns dropping by almost 50 per cent each year, as popularity for the new fad waned.

By 1922, however — by which time John Duncan had succeeded Wohlmann as government balneologist — sales of radon water had declined to fewer than 300 glasses a year, and the general manager of Tourist and Health Resorts asked Duncan to 'take steps to have the sale of radium water discontinued'.[26] The declining sales, and the manager's wish to stop offering the treatment, were possibly because of increasing evidence that radium could be harmful. Radon water sales continued for two years after the request to discontinue, but by 1925, when the general manager asked the tourist manager now in charge of the Rotorua Bathhouse about the location of the radium emanation apparatus, he was told that the jar — inside of which was 'a porcelain cylinder containing what appears to be a kind of earth' — was chipped and broken and sitting in a storeroom.[27] Hopefully the tourist manager didn't investigate the 'earth' too closely, because there are potentially fatal consequences of touching even minute quantities of radium, consequences which by this time were beginning to be appreciated.

While radium was now in widespread use, there were not well-established standards, and the nature and duration of medical treatment was left to the individual practitioner. Smaller regional hospitals continued to request funds for their own radium supplies, but by 1931 the New Zealand economy was suffering from the Great Depression. The Department of Health's budget had been slashed, and money was no longer available for what was still considered by many people to be an unproven treatment. The Department of Health's stance was that 'knowledge in regard to radium and its uses is in a state of flux and the position needs clarifying as regards dosage, technique, and the form in which the Radium should be applied before further large sums of money are spent in purchasing additional supplies of this therapeutic agent'.[28] By this time, however, the British Empire Cancer Campaign Society (the forerunner of today's Cancer Society) was active in New Zealand and began subsidising smaller centres to purchase their own stocks of radium.

HAZARDS OF RADIOACTIVITY REALISED

The degree of public enthusiasm for x-rays and radium in the first decades of the twentieth century is surprising when you consider that their ability to cause superficial burns and more serious damage was already well known. American inventor Thomas Edison had developed one of the first fluoroscopes, which produced instantaneous x-ray pictures by projecting the rays onto a fluorescent screen rather than onto a photographic plate. He and his assistants suffered problems with their eyes following x-ray exposure as early as 1896, and skin damage following x-ray exposure was reported later that year. In 1901, Pierre Curie and Henri Becquerel published a paper in which they described the burns they had received — both intentionally and inadvertently — from exposure to radioactive substances. It was public knowledge that there were dangers associated with radioactivity. In his

1905 lecture to a crowd at Canterbury University College, Rutherford had complained that he was not able to illustrate his lecture on 'Radium and its Transformations' with striking experiments 'because it was hardly safe to carry in one's cabin a sufficient amount of radium to make such experiments'.[29]

The science of radiology was new, and practitioners in New Zealand as well as overseas experimented with arrangements of equipment and procedures. As with any new technology, there were accidents and mishaps. While it was noticed very early on that x-rays could cause skin burns, the more long-term effects of the radiation were not initially known. One of the first New Zealand casualties of the new technology was Mr Wright, the electrical instructor at the Thames School of Mines, who in 1905 overexposed his hand in an x-ray machine. John Campbell, Rutherford's biographer, recounts that the 'wound would not heal, creating great medical interest as the first case of x-radiation injury in Australasia'.[30] Wright found the pain so intense that he chose to have his hand amputated rather than travel to the United Kingdom for treatment.

Medical practitioners from around the country received radiation damage to their hands, from either handling radium or from exposure to x-rays. As well as being exposed to scattered radiation, resulting from poor shielding of x-ray tubes, operators regularly exposed themselves to radiation through a process known as 'setting the tubes'. The operator placed their hand between the radiation tube and the fluorescent screen, and adjusted the equipment until a sharp image of the bones of the hand appeared on the screen. When using a screen rather than film, the exposure had to be long enough for the doctor to study the picture, and they would often stand in the path of the radiation to get a better view of the screen. Frequent and prolonged use of these practices often led to radio-dermatitis, and eventually to radiation-induced cancer. Many New Zealand medics suffering x-ray burns complained of ongoing pain, and some were forced to retire. A dentist who had seen many colleagues with radiation burns to their hands described the condition as 'excruciatingly painful . . . absolutely demanding morphia for the control of pain'.[31]

Other medics – including an Auckland Hospital radiographer; Keith Macky, an Auckland orthopaedic surgeon who did a lot of x-ray work; and Donald Goodwin of Whangarei, whose desk was next to the x-ray chest stand – would all eventually die from cancer. Radium was dangerous, too.

Christchurch's Clennell Fenwick suffered radiation damage to his fingers from handling radium, and Percy Cameron, who worked first in Dunedin and then in Wellington, suffered blindness, as well as hand and bone injuries, caused by radium handling.

Patients, too, were often victims of what Colin Anderson later referred to as 'ignorance' or 'irresponsibility'.[32] In his history of the development of radiology in New Zealand, he recounted an episode in a North Island hospital where a patient was receiving treatment to the anterior chest wall. The hospital matron who was administering the treatment was called away and forgot about the patient, who suffered an x-ray burn the thickness of his chest wall. Several years later a man died in Timaru Hospital from a huge carcinoma of the anterior abdominal wall, which had developed at the site of extensive x-ray burns received some years earlier while he was receiving radiotherapy. Dental x-rays had their own hazards. An account in the *New Zealand Dental Journal* told of a patient who suffered complete hair loss for six months, as well as persistent dermatitis of the scalp, following a series of dental x-rays.

The lack of precision (compared to today) in the early use of x-rays resulted in varying doses being received by patients, in part due to a lack of consensus as to the appropriate dose as well as imprecise practices and equipment. Decades later, Auckland radiographer John Campbell recalled the procedure for taking an ankle x-ray in the 1920s, which illustrates the potential for electrical (as well as radiation) danger posed by much of the early x-ray equipment:

> The expose time for an ankle was about 15 seconds. As no timing device was incorporated in this model we used the old photographic method — one, two, three, etc., etc., eventually arriving at 15. Another couple of seconds were added for luck. This proved alright if the tube had not emitted several sparks and frightened the patient and he or she had not moved . . . .[33]

While it was established early on that radiation could cause superficial burns and skin irritations, by 1920 a terrible irony had emerged concerning the medical use of radiation. While radiation was found to be wonderfully effective in *treating* cancer, scientists and physicians concluded it was also instrumental in *causing* cancer, as well as sterility, bone disease and other afflictions.

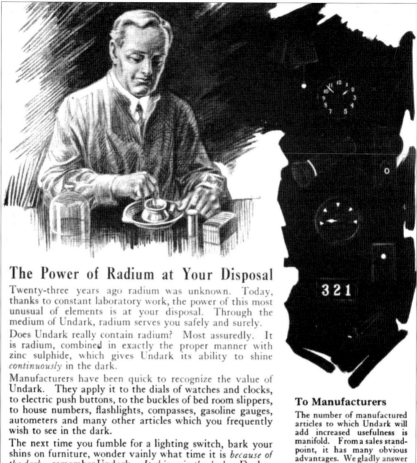

The Power of Radium at Your Disposal

Twenty-three years ago radium was unknown. Today, thanks to constant laboratory work, the power of this most unusual of elements is at your disposal. Through the medium of Undark, radium serves you safely and surely.

Does Undark really contain radium? Most assuredly. It is radium, combined in exactly the proper manner with zinc sulphide, which gives Undark its ability to shine *continuously* in the dark.

Manufacturers have been quick to recognize the value of Undark. They apply it to the dials of watches and clocks, to electric push buttons, to the buckles of bed room slippers, to house numbers, flashlights, compasses, gasoline gauges, autometers and many other articles which you frequently wish to see in the dark.

The next time you fumble for a lighting switch, bark your shins on furniture, wonder vainly what time it is *because of the dark*—remember Undark. *It shines in the dark.* Dealers can supply you with Undarked articles.

For interesting little folder telling of the production of radium and the uses of Undark address

To Manufacturers

The number of manufactured articles to which Undark will add increased usefulness is manifold. From a sales standpoint, it has many obvious advantages. We gladly answer inquiries from manufacturers and, when it seems advisable, will carry on experimental work for them. Undark may be applied either at your plant, or at our own.

The application of Undark is simple. It is furnished as a powder, which is mixed with an adhesive. The paste thus formed is painted on with a brush. It adheres firmly to any surface.

RADIUM LUMINOUS MATERIAL CORPORATION
58 PINE STREET   -   -   -   -   NEW YORK CITY
Factories: Orange, N. J.   Mines: Colorado and Utah

Radium Luminous Material

Shines in the Dark

'Undark' luminous paint, which was available in New Zealand, used radium to make things glow in the dark. This American advertisement ran in 1921; the same year the first employee of the United States Radium Corporation died after using the paint to illuminate products like wristwatches, dolls' eyes and fishing lures. Downloaded from Wiki Commons http://commons. wikimedia.org/wiki/File:Undark_(Radium_Girls)_advertisement,_1921.jpg, 18 Jan. 2012.

The cancer risk of working with even small amounts of radioactive material came to public light in the late 1920s with the widely published fate of the radium dial painters in the United States. The United States Radium Corporation factory in New Jersey employed up to 250 dial painters, mostly young women, who sat side by side at long workbenches using radium paint to illuminate numerals on the dials of soldiers' wristwatches, aeroplane instruments, and other military equipment. To get a fine point on the brush for more control over their work, the women would wipe the radium-contaminated brush between their lips. While the dangers of radiation from radium were known by this time, radium paint — containing one part radium to some 30,000 parts zinc sulphide — was not believed to be dangerous. The radium paint business flourished after the First World War, with radium used to illuminate millions of wristwatches, along with dolls' eyes, gun-sights and fishing lures.

Then, between 1921 and 1924, three of the Corporation's dial painters died, seemingly from natural causes. Many of the other dial painters began having serious problems with their teeth and jaws. The dial painters were variously diagnosed with necrosis of the jaw, phosphorus poisoning, anaemia and stomach ulcers. A 1925 investigation found the incidence of anaemia and infected mouths among former employees of the United States Radium Corporation to be beyond coincidence. More thorough investigations followed. All the dial painters examined had abnormal blood counts. On being examined in a dark room, the women were found to be luminous — their faces, hair, hands, arms, even their corsets and underwear glowed in the dark, contaminated by minute particles of radium. The same year, a paper in the *Journal of the American Medical Association* described radium's 'deadly . . . rays' and reported that once radium entered the body it spontaneously and continually irradiated the 'blood forming centres' and over time could cause severe anaemia and other disorders.[34] By 1928 at least fifteen dial painters had died from confirmed radium poisoning. Five former employees of the United States Radium Corporation received wide publicity and public sympathy when they filed a lawsuit against the company.

The injuries and deaths of the radium dial painters taught scientists and doctors that internal exposure to radium could be fatal, and had to be controlled. Continued medical investigations of the dial painters provided information about the behaviour of ionising radiation in the body, showing that, rather than being passed straight through the body as previously thought, these isotopes accumulated in various organs — radium tended to accumulate in the bones — from where they irradiated the surrounding cells. At about the same time as these links were being made, the American biologist Hermann Muller established that x-rays could induce detrimental genetic mutations and chromosomal changes in fruit flies. This 1927 discovery provided a vital clue as to how exposure to radiation could lead to cancer.

Evidence for the links between radiation exposure and cancer continued to build and was published internationally, including in New Zealand. In 1929 the American Medical Association condemned the use of radiation to remove unwanted body hair, and in 1932 it withdrew radium from its list of remedies approved for internal administration. By 1934, more than 200 American radiologists were reported to have died from cancers attributed to radiation exposure. And in July that year, Marie Curie, the celebrated discoverer of radium, died from pernicious anaemia after suffering years of ill health resulting from her exposure to radium; her fingers were already scarred by a painful radio-dermatitis, and her eyes clouded by radiation-induced cataracts.

In New Zealand, there was growing public awareness of the hazards of radiation exposure. *The Evening Post* in January 1933 reported from London on the perceived need for control of radium, which Lord Lee of Fareham described as 'the most lethal and dangerous of poisons'.[35] In 1932, the American millionaire industrialist and golf champion Eben Byers had died after drinking three bottles of radioactive water a day for several years. His anaemia, brain abscess and necrosis of the jawbone were attributed to radium poisoning. The case was widely publicised, and radon water, along with other radium-based 'health' products, declined in popularity. Newspapers reported on Byers's fate, describing the radon water as causing 'intense suffering' and a 'slow death'.[36] By this time the sale of radon water at the Rotorua Bathhouse had been discontinued, but Byers's death did not stop an enterprising company trying to sell radon water in New Zealand in the 1950s.

Some New Zealand doctors continued to use radium to treat benign medical conditions, however, despite medical evidence that it could cause skin damage, radiation burns and cancer. As late as the 1940s, x-rays were used to treat eczema, warts, acne and birthmarks; the practice was advertised as £1 for the first treatment and 5 shillings for subsequent treatments. Charles Hines, the radiographer in charge of Christchurch Hospital's radium department, described the practice of radiotherapy treatment of birthmarks (naevi), the patients usually being infants:

> A [radium] plaque is applied, with strapping, in close contact with the naevus for 12 minutes. The treatment is absolutely painless, which is a great advantage for it is best commenced at a very early age — six to eight weeks. Most naevi show a definite paling and diminution in size in two or three weeks after treatment, which is repeated at monthly intervals. Six or eight treatments are usually necessary to effect complete disappearance, though naturally, much depends on the size of the lesion, which can vary from the size of a lentil to — in some cases — an area of skin covering perhaps the whole of an arm and shoulder.[37]

Despite the mounting evidence of the dangers of radioactivity, many people still considered it to be curative. In 1936, the Department of Scientific and Industrial Research (DSIR) conducted tests on a 'specially prepared flannel said to possess radioactive properties' promoted for the treatment of various complaints.[38] Thankfully for any users of the flannel, it was found to contain little detectable radioactivity. And while radon water was taken off the menu in 1925, holidaymakers and invalids continued to flock to the 'Radium Bath' at the Rotorua Bathhouse, which was widely believed to have therapeutic radioactive qualities, even though it contained 'less radium than ordinary tap-water'.[39]

SAFETY MEASURES INTRODUCED

When it came to radiation safety, New Zealand took its cues from the rest of the world. The Department of Health took the initial lead in alerting practitioners to the hazards of working with x-rays and radium, although it was some time before concerns about patients caught up with concerns about medical staff. Departmental staff kept up to date with radiation safety measures being recommended in the United Kingdom, Australia and the United States, and, while there were no official safety standards in New Zealand, hospitals were alerted to relevant literature, with recommendations that they follow the overseas standards.

In 1921, the Department of Health drew the attention of hospital medical superintendents to the September issue of the *New Zealand*

Journal of Health and Hospitals. Alongside articles on the risk of anthrax infection from Japanese shaving brushes and the disinfectant properties of tobacco smoke was a report on the preliminary findings of the British X-Ray and Radium Protection Committee on the safety of people working in x-ray and radium departments. The dangers of overexposure to radiation and x-rays were reported as including 'visible injuries to the superficial tissues' and 'derangements of internal organs and changes in the blood'. Recommendations included advice on the use of lead screens, shields and gloves to help protect operators from radiation exposure from x-rays, and guidelines for the safe handling of radium, which it advised 'should always be manipulated with forceps or similar instruments and . . . carried from place to place in long-handled boxes lined on all sides with 1 cm of lead'.[40] The report also recommended that periodic blood tests be taken from radiation workers, in order to recognise any changes at an early stage.

People working in x-ray and radium departments were also advised to work no more than seven hours a day, five days a week, and 48 weeks a year, and on their days off should spend their time 'as much as possible out of doors'.[41] These recommendations were not always heeded. John Campbell, senior radiographer for the Auckland Hospital Board, recalled in 1952 that in the 1920s he 'was on duty every night, week end, high days and holidays, without pay or time off in lieu'.[42] The same September 1921 issue of the *New Zealand Journal of Health and Hospitals* alerted readers that exposure to radium was believed to be more dangerous than exposure to x-rays, with prolonged exposure leading to possible death from pernicious anaemia. Though how seriously the article was taken is not clear — it was followed by a story on the dangers of excessive tea-drinking.

New Zealand was an early adopter of x-ray technology, but in 1922 David Wylie, Director of the Department of Health's Hospitals Division, considered that the 'general standard of X-ray work is poor in the majority of our hospitals'.[43] Wylie called for hospital boards to keep x-ray equipment up to date, and to follow the British habit of employing and training radiographers to operate the x-ray apparatus. The following year, the number of injuries and accidents to both radiographers and patients prompted the Director-General of Health, Thomas Valintine, to tighten the conditions under which radiographers operated by issuing an edict that a radiologist should always be present during the administration of x-rays.

Despite a lack of official standards, in the 1920s operator safety began to be taken more seriously. In Dunedin, radiologist Percy Cameron began wearing gloves and a lead-leather apron and retiring to a lead-lined room to operate the x-ray table. When Christchurch Hospital installed a new x-ray therapy plant, the operator was put in an isolated room whose walls and ceilings were lined with lead. Radium used for cancer treatment was now stored in a block of lead to protect the attendants from injury.

At a 1928 meeting of the International Congress of Radiology, in Stockholm, a consortium of international groups established the International X-ray and Radium Protection Committee. The committee, which was made up of scientists and doctors, made recommendations to reduce hazards in the operation and handling of radiation technology, such as specifying the thickness of lead used to shield x-ray tubes and the walls of rooms storing radium. Their recommendations had no statutory authority, however, and were directed wholly at people who worked with x-rays, such as physicians and radiographers, with no reference to the patients receiving treatment.

It was, however, at this 1928 meeting that the idea of a safe radiation dose — or 'tolerance dose' — was first considered. Radiation was initially measured by epilation dose (the amount of radiation that would make a subject's hair fall out) or multiples or fractions of an erythema dose (the amount of radiation that would cause the skin to become red and inflamed), but these were inexact and very subjective measures. The 1928 Congress chose the roentgen (R) as the unit of x-ray measurement, and defined it as the amount of radiation needed to produce a given number of charged ions in a given amount of air. In 1934, the International X-ray and Radium Protection Committee recommended an exposure limit of 0.2 roentgen a day, or 1 roentgen a week. As an extra safety precaution, the United States adopted an exposure limit of 0.1 roentgen a day. While exposure to radiation was now recognised as potentially harmful, levels below the tolerance dose were believed to be safe and unlikely to cause permanent damage.

While New Zealand was slow to adopt radiation protection legislation, the New Zealand Department of Health continued to keep medical superintendents of all public hospitals advised of international safety advice. In a 1933 letter, the department drew attention to a revised report of the British X-Ray and Radium Protection Committee published over three issues of the *Journal of the Hospital Boards' Association*, and to 'further suggestions for the safe custody of radium and precautions to be taken when using radium and radon' contained in two Australian publications.[44] The British recommendations made detailed guidelines with regard to radium therapy, designed to protect workers from the effects of beta rays on the hands, and the effects of gamma rays on 'the internal organs, vascular and reproductive systems'.[45] Once again, the committee recommended that radiation workers have three-monthly blood tests to detect any changes at an early stage. No New Zealand regulations were drawn up, however, and initiative on the part of the medical superintendents and hospitals was required to research and act on the overseas findings and recommendations.

The Department of Health circulated revised International Recommendations for X-ray and Radium Protection in 1942. The general recommendations aimed to reduce overexposure to x-rays and radium by providing staff with adequate protection — such as lead screening — and suitable working conditions. With the tolerance dose still at 1 roentgen per week, it was recommended that staff carry photographic film badges to measure cumulative radiation exposure. The international recommendations also advised that discretion should be exercised in transmitting radium salts by post, with quantities of up to 50 milligrams of radium able to be posted provided they were in lead-lined boxes.

Despite the new focus on safety, there were still mishaps and carelessness. Radium, expensive and potentially dangerous, was not subject to any standard protocols, and was not always treated with the care it deserved. On at least two occasions radium needles used in New Zealand hospitals were accidentally thrown in the rubbish and incinerated, although in both recorded accounts, from 1924 and 1939, the valuable substance was safely recovered. Well into the 1940s, however, New Zealand's medical radiation safety measures were lacking. Colin Alexander, a house surgeon at Auckland Hospital, later recalled 'the cavalier attitude to X-ray safety, notwithstanding the plain evidence of the potential dangers. . . . In Auckland, the main X-ray room had about four tables, all working simultaneously, separated only by curtains, resulting in a bath of whole body radiation for all concerned. We all had regular blood counts and had extra holidays to compensate for the hazard, but the documented lack of any detectable ill-effect eventually led to our losing the latter perk.'[46]

The New Zealand branch of the British Empire Cancer Campaign Society was formed in 1929 to encourage the establishment of cancer treatment clinics and research into the causes of cancer. While tuberculosis was the nineteenth century's biggest killer, better living conditions and hygiene standards, with a subsequent increased life expectancy, had led heart disease and cancer to eclipse tuberculosis in causing fatalities. By 1928, New Zealand's cancer death rate was 9.87 per 10,000 people per annum: nearly double the rate of tuberculosis, and second only to heart disease in causing mortality.

The society received subsidies from the Department of Health, and money from public donations. In 1933, the Travis Trust, which was established from the estate of the late William Henry Travis of Christchurch, offered the society £500 a year for three years to pay the salary of a physicist. Jack Strong — a Victoria University College physics graduate and a radium attendant at the hospital since 1932 — was appointed to run the laboratory. After training in Australia, Strong set up a laboratory and workshop in the basement of Wellington Hospital, where he took charge of the radium plant previously staffed part time by science students, and began to help cancer treatment centres to deal with the physical aspects of radiation therapy. As well as operating the radon plant — which captured the radon gas emanating from a supply of radium for treatment centres around New Zealand — he set up New Zealand's first standards for measuring radiation dosage, and travelled to the four main centres to provide advice to radiotherapists and to calibrate x-ray apparatus used in radiotherapy. It was now recognised that the success of radiotherapy depended on the accurate measurement and administration of the treatment dose: successful treatment was a function of experienced technicians and well-calibrated equipment.

Strong found working conditions at Wellington Hospital difficult — not least because the basement room used for standardisation work had high levels of radioactive contamination which affected results — and in 1937 the society moved its radiological service to Christchurch. The Travis Radiophysical Laboratory was transferred to the science department of Canterbury University College, where Strong was joined in 1938 by George Roth (who succeeded Strong when he died in 1941).

The laboratory's first task was to set up standardisation equipment, and then to calibrate all of New Zealand's therapeutic x-ray plants and clinical dosimeters (used to measure radiation dose) to this standard. Most developed countries, including the United Kingdom, the United States and Germany, inspected x-ray installations only as required and on a cash-for-service basis. In contrast, New Zealand, like Sweden and Australia, established a nationwide physical calibration scheme whose service was free of charge.

The New Zealand service was unusual in that it was set up and organised by a private body, the British Empire Cancer Campaign Society, under close co-operation with the Health Department. For the next ten years, the Travis Radiophysical Laboratory provided a free service whereby its two physicists visited up to 26 x-ray therapy plants and 18 clinical dosimeters twice a year to calibrate equipment against the laboratory's portable standards and to advise on measures to protect patients and staff from radiation and electrical risks.

In the 1940s, radiologists began taking a leading role in promoting the safe use of radiological equipment. Despite the safety measures introduced by individual hospitals during the preceding two decades, by the 1940s some New Zealand radiologists were expressing concern about the safety of their staff and patients. With no national standards or monitoring procedures in place, it fell to individual radiologists to ask the laboratory to measure the cumulative dose received by individual staff members, and to advise on improvements to existing protective measures. It had also become clear that, without a complete record of all publicly and privately owned x-ray plants, it was impossible to ensure adequate protection for all x-ray workers at the country's many hundred widely scattered installations. The laboratory was impressed by the radiologists' requests, and concerned by the 'severe damages' which they observed had been sustained by some workers.[47] After discussions at a 1943 conference of radiologists, recommendations were made for regulations to cover electrical and radiation protection. The next year, in the interest of safety and accuracy, New Zealand radiologists called for all users of x-ray equipment to be registered, and for calibration visits to be extended to all diagnostic x-ray plants, as well as to therapy plants.

The Electrical Wiring (X-ray) Regulations were introduced in 1944, by which time there were at least 450 x-ray plants believed to be operating in New Zealand. These regulations provided technical provisions for the electrical safety of x-ray plants, and for the first time required the registration of all x-ray plants. It was not until 1949, however, that radiation protection was also subject to regulation.

By the beginning of the Second World War, the New Zealand medical system was dependent on radiation technology. Many of the staff now working in radiology departments had studied overseas, and radiology and radiography were established professions. The free hospital care established under the first Labour Government's Social Security Act 1938 included

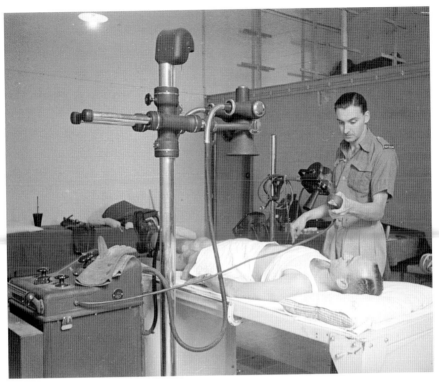

The Second World War led to increased demand for x-rays at home, and better training for radiographers who worked in Europe. This photograph shows the x-ray department of the New Zealand General Hospital in Molfetta, Italy, in 1944. Photograph by George Bull, DA-05841-F, PAColl-4161, Alexander Turnbull Library, Wellington, New Zealand.

x-ray examinations and treatment, which led to an increase in demand for diagnostic x-rays and an increased workload for radiographers and radiologists. Demand increased further when out-patient benefits were introduced in 1941 and many people took the opportunity to have a free medical check-up.

The start of the Second World War led to a further increase in the demand for x-rays. In September 1939, as a step towards eliminating from the armed services people suffering from active tuberculosis, x-ray plants were installed in the military training camps at Burnham, Trentham and Papakura, and all military recruits were given a chest x-ray. By the 1940s the death rate from tuberculosis was at an all-time low, but it was still a significant health problem, particularly for Maori, whose tuberculosis death rate was up to ten times higher than that for non-Maori. In 1941, the Department of Health began x-raying every Maori secondary school pupil to try to stop the spread of undiagnosed tuberculosis. The establishment of the Department of Health's new Division of Tuberculosis further increased the demand for x-rays. Their research into the incidence of tuberculosis involved the initiation of mass miniature x-ray programmes for early identification of the disease. Chest x-rays were required for all new apprentices, for entrants to teachers training colleges, and for trainee nurses and dental nurses. (Applicants found to have tuberculosis were not considered acceptable candidates.) This all put a strain on the country's x-ray facilities, and hospitals were unable to keep up with demand.

THE BIRTH OF NUCLEAR SCIENCE

While scientists and medics throughout the world were taking advantage of the discoveries of x-rays, radium and radioactivity, physicists and chemists continued to seek a better understanding of the atom, its particles, and their properties. In 1908, Rutherford was awarded the Nobel Prize in Chemistry for his work on radioactivity. He was now based at the University of Manchester and the following year, with the help of his assistant Hans Geiger and a 20-year old student called Ernest Marsden, Rutherford worked on a series of experiments in which a beam of alpha particles was scattered after passing through a thin metal foil. In response to Geiger's advice that Marsden was ready for a research project of his own, Rutherford asked Marsden to see if he could get evidence of alpha particles directly reflected from a metal surface. In a now-famous experiment, Marsden observed that, instead of passing through, a tiny fraction of alpha particles were deflected straight back from a thin gold foil. Rutherford later described this result as being 'almost as incredible as if you had fired a fifteen-inch shell at a piece of tissue paper and it came back and hit you'.[48]

Two years later, Rutherford came up with a revolutionary new model for the structure of the atom, with a centralised concentration of mass and positive charge — which he called the nucleus — surrounded by empty space and a sea of orbiting negatively charged electrons.

In a 1903 paper on 'radioactive change', Rutherford and Soddy had concluded that 'the energy latent in the atom must be enormous',[49] and speculated about the release of energy from the atom and the possibility

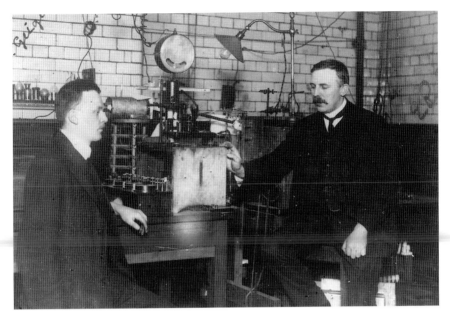

Hans Geiger — whose name was given to what we now call a Geiger counter — was Ernest Rutherford's assistant in his laboratory at the University of Manchester. Sir Ernest Marsden Collection, F-65890-1/2, Alexander Turnbull Library, Wellington, New Zealand.

of atomic weapons, with Rutherford jokingly speculating that 'some fool in a laboratory might blow up the universe unawares'.[50] Against general scientific opinion, the novelist H. G. Wells predicted in 1914 that the energy in the atom would become available for use. In *The World Set Free*, Wells envisaged a European conflict involving 'atomic bombs' and after which 'atomic energy' would be used for the benefit of humankind. The book was widely discussed, but Rutherford dismissed Wells's claims, saying to *The New Zealand Herald* in 1914 that 'it did not appear within the region of possibility that a substance could be manufactured having the properties of Wells's bombs'.[51]

Rutherford left Manchester in 1919 and returned to the Cavendish Laboratory at Cambridge, this time as director of the laboratory. There, from a series of experiments in which he had bombarded nitrogen atoms with alpha particles from a radium source, he concluded that the nitrogen atoms were being transformed into oxygen atoms, with a hydrogen nucleus (later called a proton) ejected. Rutherford had 'split the atom', and the science of nuclear physics — the manipulation of the atomic nucleus — was launched.

New Zealand's Ernest Rutherford (standing, fourth from right, with big moustache) was at the forefront of international physics. Rutherford attended the 1911 Solvay Conference in Brussels, along with other physics pioneers such as Marie Curie, Albert Einstein and Max Planck. PAColl-6238-32, Alexander Turnbull Library, Wellington, New Zealand.

Under Rutherford's instruction, Cavendish Laboratory physicists John Cockcroft and Ernest Walton spent years designing and building electrical devices to accelerate streams of electrons and protons. In 1932, using a handmade array of transformers, rectifiers, glass tubes and vacuum pumps, Cockcroft and Walton bombarded a target of lithium with high-voltage protons and detected alpha particles on a scintillation screen, evidence that some of the lithium atoms had absorbed a proton and split in half. While Rutherford had used a natural source of radioactivity to knock a proton out of a nitrogen atom's nucleus, Cockcroft and Walton had literally split a lithium atom in half. ('The atom split, but world still safe,' read the *Sunday Express*. 'Let it be split, so long as it does not explode,' said the *Daily Mirror*.)[52] Cockcroft and Walton's splitting of the lithium atom was artificial disintegration by artificial means, and they soon completed experiments into the energy releases and mass changes to confirm Albert Einstein's equation that $E = mc^2$ (energy equals mass times the square of the speed of light).

As the world was poised to decide how to deal with new forms of nuclear technology, in the form of nuclear weapons and nuclear energy using uranium as fuel, New Zealand's role would be cemented by the presence of Rutherford's former student Ernest Marsden as head of the country's

As head of the Department of Scientific and Industrial Research from 1926 to 1947, Ernest Marsden championed the development of nuclear sciences in New Zealand. Sir Charles Fleming Collection, 1/4-018564-F, Alexander Turnbull Library, Wellington, New Zealand.

State science organisation. Marsden had come to New Zealand in 1915 to take, at Rutherford's recommendation, the position of professor of physics at Victoria University College in Wellington. In 1926, he was appointed first permanent secretary of New Zealand's brand-new Department of Scientific and Industrial Research (DSIR). In this role, Marsden became a champion of nuclear science and technology in New Zealand. As well as dealing with the management and administration of the DSIR — whose initial focus was on scientific issues of relevance to agriculture — Marsden continued to be involved in his own research and, although he was scientifically isolated in New Zealand, he kept up to date with international developments in the world of nuclear physics and was quick to adapt new technologies for New Zealand. Following Victor Hess's discovery of cosmic radiation, Marsden measured cosmic radiation levels in Apia (Samoa) and on Mount Egmont in Taranaki, and later established a cosmic-ray meter at the DSIR's Magnetic

Observatory in Christchurch, which sent its results to the Carnegie Institute in Washington for compilation with other results from around the world. Marsden also pioneered the non-medical use of artificially radioactive isotopes in New Zealand, in innovative agricultural experiments into the role played by cobalt in animal metabolism. In the late 1930s, working with a young DSIR scientist called Charles Watson-Munro, Marsden conducted a survey of radioactivity in New Zealand soils, in an attempt to establish a connection between radioactivity and the regional incidence of goitre.

Other New Zealand scientists had examined the radioactivity of New Zealand soil and water through the 1920s. Particular attention was paid to geothermal water and to Christchurch's artesian water supply, which was found to have a radon content ten to 20 times higher than other New Zealand samples tested. Other researchers made use of x-rays for experiments such as chemical and crystal analysis, or as a remote means of detecting mouldy apple cores in apples destined for export markets.

In 1933 Rutherford, who had launched the field of nuclear physics and earned himself the title of the 'father of nuclear physics', reiterated his belief that the power of the atom could not be harnessed, saying that the 'energy produced by the breaking down of the atom is a very poor kind of thing. Anyone who expects a source of power from the transformation of these atoms is talking moonshine.'[53] But by the dawn of the Second World War, the understanding of the atom was growing in complexity, and, true to H. G. Wells's predictions, so was the possibility of a future when atomic transmutation would provide the means of providing nuclear power and destructive weapons. Rutherford, however, was not to see the application of this science. Lord Rutherford of Nelson — as he had elected to be known when he was elevated to the peerage in 1931 — died in October 1937, just two years before the start of the Second World War set in train developments that would soon lead to the manufacture and use of nuclear weapons and energy. It was Rutherford's student, Ernest Marsden, as head of New Zealand's DSIR, who would now have the greatest influence on New Zealand's involvement in the new field of nuclear science.

2

★

Some fool in a laboratory
The atom bomb and the dawn of the atomic age

How proud New Zealand must be that the foundations of the
amazing discovery concerning latent atomic energy were laid by
her own great scientist Rutherford. — VISCOUNT BLEDISLOE
IN TELEGRAM TO NEW ZEALAND, 9 AUGUST 1945[1]

When the first atomic bomb exploded, the world as we
have come to know it came, I believe, to an end.
— KARL POPPER, CHRISTCHURCH, AUGUST 1945[2]

After the atomic* bomb was dropped on Nagasaki, Japan, Viscount
Bledisloe, New Zealand's former Governor-General, congratulated
the country on her role in the victory: Rutherford's work on the
atomic nucleus was acknowledged as laying the scientific foundations for the
development of what was then known as the 'atomic bomb'.

Rutherford was not the only New Zealand scientist involved in the
victory. After the bomb was dropped, the Government revealed the formerly
secret role of a group of New Zealand scientists who had worked on the
bomb programme, and told New Zealanders that they should be 'proud to

---

* The terms 'atomic' and 'nuclear' are used interchangably in this book — while 'nuclear' is more
  correct when referring to energy and weapons, the word 'atomic' was used more commonly in the
  mid-twentieth century.

know that some of her scientists of this generation were at the forefront of this latest development'.[3]

The fact that some New Zealand scientists had been involved at all was due, in good part, to New Zealand's Rutherford connection, and the efforts of Rutherford's former student Ernest Marsden. Thanks to Marsden, a team of young New Zealand scientists worked on the Manhattan Project, the American-led project to develop the first nuclear weapons, and on the Canada-based British-led project to develop nuclear energy. While the New Zealanders' role in the massive Manhattan Project was minor, the New Zealand team in Canada played a significant role in the development of the first nuclear reactors in Canada and subsequently in the United Kingdom.

PHYSICS AND THE SECOND WORLD WAR

New Zealand had joined Britain in declaring war on Germany in September 1939, with New Zealand's strong links with Britain demonstrated in Prime Minister Michael Joseph Savage's loyal declaration that 'where she goes, we go, where she stands, we stand'.[4] Over the next six years, as the war escalated and expanded, more than 200,000 New Zealanders would serve in the war in Europe and the Pacific. Following the outbreak of the Second World War, Marsden, in his position as secretary of the DSIR, was given the title of Scientific Advisor to the Defence Department and, later, Director of Scientific Developments, in which role he was charged with mobilising New Zealand's scientific manpower. Conscription to the armed forces was introduced in 1940 and, after 1942, remaining workers, including scientists and science students, could also be directed by Manpower Committees into what were considered essential industries.

The DSIR put its efforts into supporting the war effort, particularly through advances in agriculture and food science, and by finding local sources of scarce goods. Geologists were directed to find strategic minerals like mica and asbestos, and chemists experimented with new fuel sources and created insect- and shark-repellents for soldiers stationed in the Pacific. The DSIR's biggest efforts, however, were in the physical sciences — Marsden's field of expertise — with the establishment of two new physical sciences laboratories that grew to become two of the DSIR's largest divisions. Scientists at the Wellington-based Radio Development Laboratory were involved in a secret programme to develop radar in New Zealand, which was initially the Allies' top scientific priority. The Physical Testing Laboratory, which in 1943 was renamed the Dominion Physical Laboratory, was established to cater to the armed services' demands for physical testing and calibration of instruments.

The war also saw radium use in New Zealand expand beyond medicine. In January 1944, a 'luminising laboratory' was set up in a Dominion Physical Laboratory base in Lower Hutt. Four young women were employed to apply radium-activated luminous paint to dials and markings on radio communication equipment for the New Zealand armed forces to use in jungle and tropical areas. They were only one group of workers using radioactive paint during the war. In 1943, the Director-General of Health wrote to Marsden about the use of radioactive paints at a factory in Auckland and in an Air Force factory, expressing his concern that 'workers do not realise the tremendous danger of the inhalation and ingestion of the powder used'.[5] Travis Radiophysical Laboratory physicists advised Marsden that regulations should be made to cover every person handling unsealed radioactive substances, with a six-monthly radon-exhalation test and a Geiger-counter test. Marsden, however, was not worried about the workers' safety. He thought that regulations would be premature and the cost 'excessive for the small amount of work to be done which is only on an experimental scale'.[6] It was against this background, and in the absence of any New Zealand regulations, that the DSIR laboratory was established. The laboratory was set up, however, with reference to the British Factories (Luminising) (Health and Safety Precautions) Order 1942 and was required to comply with British standards as modified by the Travis Radiophysical Laboratory. Copies of the United States Bureau of Standards Handbook on Safe Handling of Radioactive Luminous Compound were also circulated by the Department of Health.

Despite the well-publicised plight of the American radium dial painters in the 1920s, safety was not given the priority it deserved and conditions in the Lower Hutt laboratory were not up to international standards. The young women employed by the laboratory worked in a makeshift laboratory set up in three small Army huts. They sat on upholstered seats above floors covered in tarred paper, in front of Vitrolite-covered benches below a hood with an exhaust draught to remove radon gas and stray particles. They applied paint with brushes, and wore gowns, aprons, caps and gloves, washing the gloves at the end of the day.

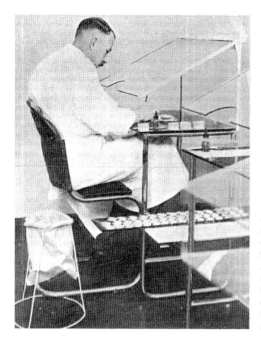

After the review of the laboratory's safety measures, staff worked in a ventilated painting cabinet, with hard surfaces that were easy to clean. This photograph is probably of J. M. C. Tingey, who led the Dominion Physical Laboratory team that used radioactive paint to illuminate dials and ammeters for military use. J. M. C. Tingey, 'Luminizing Army Radio Equipment', *New Zealand Journal of Science and Technology*, 27B, 1945, p. 140.

But when the luminising laboratories were inspected in May 1944, conditions were found wanting. The makeshift laboratory was condemned as 'ill-adapted for this purpose [making] it unnecessarily difficult to maintain a high standard of hygiene and safety'.[7] Radioactive contamination was found on various tins and bottles in cupboards, in crevices around painting hoods, and even in a bottle of sweets that had been surreptitiously introduced into the painting room. The gloves the women washed at the end of the day and hung up to dry were found to be 'heavily contaminated with luminous paint'.[8]

The Dominion Physical Laboratory was receptive to the critical report, and the laboratory was closed for three weeks while they wet-cleaned the painting room, installed new work stations, laid linoleum on the floor, replaced upholstered chairs with easily-cleaned steel tube chairs, installed new ventilation systems, and replaced paint brushes with resin-impregnated wooden meat skewers that were disposed of after each working period. From September 1944, breath samples were collected from each staff member of the luminising laboratory and sent monthly to the radiophysical laboratory in Christchurch, where physicists had developed equipment for measuring the radon content of operators' breath to detect evidence of any ingested or inhaled radium. In December 1944, results of the radon exhalation tests revealed that four of the five staff had results between 10 and 80 per cent of the tolerance dose (the acceptable limit) and one had a result of 130 per cent of tolerance dose, which reduced to less than 10 per cent of tolerance after a rest and an unspecified treatment. George Roth stated that the radon exhalation tests 'proved that the protective measures recommended for luminising work in New Zealand were wholly successful in that all the workers examined at the end of their employment remained well within the tolerance limit for ingested and inhaled radium'.[9]

UNLEASHING THE ENERGY OF THE NUCLEUS

While radium and x-rays had been the great discoveries in physics at the end of the nineteenth century, early twentieth-century developments in the new physics were now culminating in a wartime project to unleash the energy inside the atomic nucleus, thanks in part to the work of Ernest Rutherford. The first explorations into the nucleus had been made by Rutherford in 1909 when he had devised the experiment that led to his discovery of the nucleus. Then, in 1917, Rutherford had 'split' a nitrogen atom, and in 1932 his team halved a lithium nucleus.

A new technique was now needed to investigate the nuclei of heavier elements. The Italian scientist Enrico Fermi realised that slowing the neutrons, through collisions with water or another hydrogen-containing substance, improved experiment outcomes. So Fermi began bombarding heavy elements like uranium — the heaviest naturally occurring element — with these slow neutrons, and sometimes, and somehow, began creating new elements heavier than those he had started with. (Some of the atoms he was working with were absorbing the neutrons.) German chemist Otto Hahn and physicist Lise Meitner were inspired by Fermi's work and set out to repeat his experiments. After Meitner fled Nazi Germany for Sweden, Hahn continued the work with a colleague, Fritz Strassmann, conveying the results to Meitner by post. In Sweden, with the help of her physicist nephew Otto Frisch, Meitner realised that by bombarding uranium with slow neutrons Hahn had fractured atoms of uranium into lighter elements, such as barium, with more neutrons, and energy, being released in the process. Meitner then calculated the incredible release of energy from the disintegration of uranium, which Frisch called 'fission', as being in the order of 200 million electron volts from a single atom. In principle, given a great enough mass (a *critical mass*) of a heavy material such as uranium, the neutrons emitted by the fission of one atom would initiate fission in neighbouring atoms, setting off a chain reaction and initiating an incredible release of energy. When these results were published in 1939, a global search began for a way to harness this energy, and, with war imminent, the focus was on using the energy for a bomb.

Scientists in the United Kingdom, the United States and the Soviet Union soon began working on fission, but by the start of the Second World War Germany was the only country to have a military office focused on nuclear energy. When, in 1940, it was established that uranium-235, an isotope comprising only 0.7 per cent of natural uranium, was responsible for fission, scientists began to focus on separating uranium-235 from natural uranium, which contains mostly uranium-238, and on determining the critical mass that would be needed to sustain a chain reaction.

In 1940, Britain set up the MAUD Committee, which focused on directing secret research towards producing a uranium bomb. Four of the five original committee members were Cavendish Laboratory alumni who had studied under or worked with Ernest Rutherford: James Chadwick, John Cockcroft, Philip Moon and the Australian physicist Mark Oliphant. The MAUD Committee's reports, presented in June and July 1941, recommended that it was feasible to make a uranium bomb with as little as 25 pounds (about 11 kilograms) of uranium-235, and that it would be possible to create electricity from nuclear energy, using uranium as the fuel source and heavy water* as a moderator to slow the neutrons. Their report stated that a 25-pound uranium bomb would not only have the equivalent destructive power of 1800 tons of TNT, it would 'release large quantities of radioactive substances, which would make places near to where the bomb exploded dangerous to human life for a long period'.[10] When the British Government accepted the MAUD Committee's recommendations to proceed with both a nuclear power project and a nuclear bomb project, a new directorate with the deliberately nonsensical code name Tube Alloys was established within the British DSIR. Following the MAUD Committee's advice to secure

---

* Heavy water is a form of water (H_2O) that has a heavy hydrogen atom — a hydrogen atom with a neutron and a proton in its nucleus, called deuterium — in place of a regular hydrogen atom, the nucleus of which contains only one proton.

control of uranium supplies, Britain initiated a Commonwealth search for uranium, but felt secure in the knowledge that of the world's two largest known uranium supplies — in Canada and in the Belgian Congo — one was on Commonwealth land.

The United Kingdom's nuclear programme was initially more advanced than the American programme, where scientists were also working on nuclear projects. There was, for a time, a free exchange of information and ideas between scientists from the two countries, although this was mostly from the United Kingdom to the United States. Following the American entry into the Second World War in 1941, however, the United States launched a co-ordinated and well-financed effort to develop nuclear bombs. One of the first key sites was at Oak Ridge, Tennessee, where a plant was built to separate uranium-235 from uranium-238 using both electromagnetic and gaseous diffusion techniques. At Hanford, Washington, a nuclear reactor was built to produce a newly discovered artificial element, plutonium, as an alternative fuel source for a fission weapon. At Los Alamos, in New Mexico, scientists were tasked with putting the raw materials together to design and produce a nuclear bomb. Together, these sites formed the Manhattan Project, which became the world's biggest-ever scientific endeavour.

When it was clear that the American nuclear project had advanced well ahead of the British project, the United Kingdom sought to join forces with the United States, who had stopped sharing information. After the signing of the Quebec Agreement of August 1943, under which the British and American scientists would collaborate on their nuclear energy projects, the Tube Alloys Project was subsumed into the American nuclear programme and groups of British scientists began working on specific aspects of the Manhattan Project. Oliphant went to Berkeley in November 1943, to work with his friend Ernest Lawrence, the inventor of the cyclotron, on the separation of uranium-235 by electromagnetic means. Two other Australian scientists, Harrie Massey and Eric Burhop, also Cavendish Laboratory graduates, joined the British group at Berkeley, where Massey led a group of theoretical physicists.

NEW ZEALAND SCIENTISTS IN NORTH AMERICA

It seems surprising that of all the other Commonwealth countries, it was New Zealand — a small country now known for its 'nuclear-free' status — that played a significant role in the British nuclear programme in Canada. But, as historian Ross Galbreath has pointed out, it makes sense when New Zealand's involvement is recognised as being primarily due to the efforts of Ernest Marsden, and the result of New Zealand's Rutherford connection to nuclear science. In his role as head of the DSIR, Marsden made several wartime trips to the United Kingdom, mostly to advance the secret programme to develop radar in New Zealand. There, he learnt of the nuclear programme through his many contacts in the British physics community.

In December 1943, Marsden was travelling through the United States on his way to the United Kingdom when, in Washington DC, he chanced upon James Chadwick, by now scientific director of the British nuclear research project, as well as Australian physicist Mark Oliphant and Danish physicist Niels Bohr, who had been smuggled out of Denmark and was travelling under an assumed name. Following the signing of the Quebec Agreement, Chadwick and Oliphant were in Washington with the top-secret task of arranging details of scientific co-operation between the United Kingdom's and United States' nuclear research programmes. Oliphant later recalled that they were in their hotel lobby waiting for the elevator when they felt taps on their shoulders and turned to find Marsden in full military uniform. They were taken aback to hear Marsden say, 'I can guess why two nuclear physicists are here!'[11] During the elevator journey Marsden put in a good word for New Zealand's participation in the bomb project. He followed this up in London with Sir John Anderson, Chancellor of the Exchequer and the British minister in charge of atomic energy matters. Many of the Commonwealth scientists working on the British nuclear research programme were, like Marsden, past students or colleagues of Rutherford, and Marsden was able to trade successfully on his reputation of being involved in the birth of nuclear physics, which, as Massey later said, had earned Marsden 'a place among the immortals'.[12]

Following the necessary protocol, the British Government asked New Zealand Prime Minister Peter Fraser for five New Zealand men to join the British nuclear research team. Robin Williams, a young physicist with the DSIR's Radio Development Laboratory, recalled reporting to Wellington in July 1944 to find Marsden 'cock-a-hoop about the fact that he had managed to get a number of New Zealanders in on the atom bomb project'.[13] Young DSIR scientists George Page, Bill Young and Charles Watson-Munro joined Williams. Their terms of employment seconded them to the United Kingdom

DSIR for one year, or for the duration of the war, whichever was longer. Marsden was very keen for New Zealand to launch an atomic research programme when the war finished, and following the secondment the men were required to return to New Zealand for at least one year.

Most of the British scientists working on the electromagnetic separation of uranium had now transferred to the University of California at Berkeley. When Williams and Page joined this team in July 1944, they took the number of British men on this part of the project to 35, though Margaret Gowing, official historian of the British nuclear project, points out that 'the influence of the British was far higher than their numbers would suggest'.[14] As well as Williams and Page, there were two other New Zealand-born scientists on the team who had arrived from the United Kingdom with the British group, one being Maurice Wilkins.

The electromagnetic separation process that the British scientists at Berkeley were working on involved first accelerating ionised uranium using an electric field, then passing the beam of accelerated ions through a magnetic field which deflected the uranium-235 ions slightly more than it deflected the uranium-238 ions (because of their lower mass), and allowed for separate collection of the two isotopes. The challenge was to design and build the most efficient plant possible, and theoretical and experimental physicists were needed to help solve problems arising from the design challenge and the operation of the plant. Williams mostly worked under Massey with a group of theoretical physicists who contributed to the project by improving the team's understanding of the fundamental processes involved in uranium separation. Page, along with the engineers on the project, made significant contributions to improving and simplifying the design of the electromagnetic separation plant. Both Williams and Page, although based at the University of California in Berkeley, made several trips to the base at Oak Ridge, Tennessee, which Williams described as being like a 'workers' camp and a prison camp combined with lots of fences . . . lots of mud and prefabricated houses'.[15]

In Canada, a team of mostly English and Canadian scientists, led by another Rutherford old-boy, John Cockcroft, had begun a project to build an experimental pile, or reactor, using uranium rods in a heavy-water moderator to produce an intended energy output of 10,000 kilowatts. A rural site near the village of Chalk River, on the south bank of the Ottawa River, was chosen for the laboratory. In assembling his team, Cockcroft was seeking 'engineers with a decent physics background and physicists with a flair for

Ernest Walton (left) and John Cockcroft (right) were students of Rutherford's at the Cavendish Laboratory. When Cockcroft became head of the Canadian atomic energy project at Chalk River, and later director of the British Atomic Energy Research Establishment at Harwell, New Zealand's Rutherford connection gained New Zealand scientists an advantage. That and the fact that Marsden was known to send Cockcroft a leg of New Zealand lamb in time for Christmas. UK Atomic Energy Authority, courtesy AIP Emilio Segre Visual Archives.

engineering'.[16] Watson-Munro and Young travelled to Montreal from New Zealand; and another New Zealander, Ken George, reported directly to Montreal from his post as the DSIR's scientific liaison officer in Washington, where, in recognition of the growing importance of the relationship with the United States, New Zealand had opened a legation only two years before.

As part of the Canadian team — which comprised 40 Canadians, 40 British, and a small group of men from France and other nations — the New Zealanders began work on building a low-energy atomic pile, using natural uranium fuel and heavy water as a moderator. The leaders of the British nuclear programme were very pleased with the New Zealand scientists, and Oliphant told New Zealand's scientific liaison officer in London that New Zealand was being compared favourably with Australia for offering 'five good men without questioning the soundness of the purpose and the good faith of the British government'.[17] It wasn't really a matter of New Zealand's unquestioning support, however. Marsden strongly believed that it would be to New Zealand's advantage to have men gaining scientific experience on the world's most ambitious scientific project. Once the project was complete, they could bring this knowledge home to launch a nuclear research programme in New Zealand.

After connecting the New Zealand physicists with the nuclear project, what was Marsden's role? He could have left the New Zealand scientists to their work and focused on his many other duties as head of the DSIR, but he was preoccupied with the work in North America. As a scientist-turned-administrator, Marsden was tremendously excited about these new applications of nuclear physics and felt stymied and frustrated in his administrative and managerial role so far away from the action. He wrote regularly to the American-based scientists, asking, sometimes inappropriately, for details of their research.

In January 1945, after the New Zealand team had been in North America only a few months, Cockcroft wrote to Marsden requesting three more scientists from New Zealand. Marsden, in turn, wrote to the Minister of Scientific and Industrial Research seeking permission to send three more men to North America. He also recommended that 'in view of the overwhelming importance of the work to the future of New Zealand . . . it is in my opinion most necessary that we should have a team of men on this work and one way to do this would be for Cabinet to give authority for such a section or team and appoint the men to it to return for duty in New Zealand as soon as hostilities cease'. Marsden was either prescient in his thinking — or well informed — when he added 'I presume the full story of the operations of the T. A. project will "break" within say six months, and if we are able to announce that the Government has taken appropriate action in the matter, i.e. such as the above, it will be a source of justifiable satisfaction to the country'.[18] It was a little less than eight months later that the operations of the Tube Alloys Project 'broke', and the atomic bomb was dropped on Hiroshima.

Marsden gained permission to send the additional New Zealand scientists, and in March 1945 Arnold Allan and Gordon Fergusson, both assistant physicists with the Radio Development Laboratory, left New Zealand to join Watson-Munro's team in Canada. In response to Cockcroft's request for a third man, Marsden first recommended Ian Walker 'an excellent New Zealand electronics man in [the] UK', but as an alternative asked 'can you find room for me at a senior scientific officer grade — in any direction of work, for any period of time'.[19] At receiving no response to this offer,

Marsden repeated it to Watson-Munro, writing: 'I would be quite glad to get back to pure research for the rest of my life'.[20] Cockcroft, perhaps alarmed by Marsden's inappropriate offer, informed Marsden in May 1945 that a third man was not required — for now.

As he was unable to be involved in the North American research programme, Marsden directed his enthusiasm to plans for a nuclear research team in New Zealand after the war and a search for uranium in the South Island. In a letter to Ken George in April 1945, Marsden wrote 'we shall have a self-contained team on TA in New Zealand in due course' and 'we are having quite a lot of fun chasing radioactive minerals (don't repeat this!). They are fairly widespread in small concentrations and the problem is in care and methods of concentration.'[21] In July 1945 he gained Cabinet approval to place all the men working on the nuclear project in America, together with some remaining in New Zealand, in a special team and on the permanent staff of the DSIR.

WARTIME URANIUM SURVEY

With his characteristic positive attitude — which his colleagues later described as 'infectious enthusiasm' and 'irrepressible optimism' — Marsden had overstated the results of the uranium survey.[22] When the United Kingdom had initiated a Commonwealth search for uranium in 1942, they had excluded New Zealand, whose geology was not considered promising. Marsden, however, thought he knew better, and in December 1943, while on his fruitful trip through the United States, had taken matters into his own hands, instructing New Zealand's Geological Survey to start a search for radioactive minerals in the South Island. At about the same time, Marsden submitted for publication a radioactivity survey that he and Watson-Munro had made in the 1930s. When he was back in New Zealand, Marsden progressed the uranium survey, advising the Minister of Scientific and Industrial Research that, although the likelihood of finding commercial quantities of uranium in New Zealand was small, 'in view of the coming importance of the subject and likely demand, I recommend that authority be given for a thorough survey . . .'.[23] The New Zealand War Cabinet approved funding, and in the winter of 1944 a team of DSIR physicists assembled in Wellington to start work on the uranium project.

After reading relevant books and journals, the team decided the best way to detect uranium was to use a Geiger counter, a radiation-measuring device developed by Rutherford's assistant Hans Geiger, so they set about trying to make them, and reducing them to a reasonably portable size. By springtime, they were ready, and a mining engineer and a physicist, carrying a Geiger counter to measure radioactivity, began secretly exploring beach sands along the West Coast of the South Island, from Karamea to the Moeraki River. Surveys of Stewart Island beach and river sands, and of beach sands and dredge tailings at Gillespies Beach, followed. Later in the year, Marsden made requests to Ian Coop, the DSIR's scientific liaison officer in London, for instruments and materials for the uranium search. Coop was able to procure some items — lead, tungsten wire and sieves — and send them to Marsden in New Zealand, but, because of the war, many of the requested items were in short supply. In his reply, Coop advised Marsden that the British directorate of Tube Alloys were surprised to find out about the New Zealand uranium survey and were concerned to emphasise that Marsden must treat the project as secret.

While the United Kingdom had earlier written-off New Zealand as not being a likely source of uranium, the United States was interested in New Zealand's ability to supply nuclear projects with uranium. A Mr G. H. Hall, a representative of the American chemical company Union Carbide, which was working in close contact with the United States Government, visited New Zealand in late 1944 to assess the probable sources of radioactive minerals such as uranium and thorium. Hall reported favourably on the possibility of the cheap production of uranium oxide and thorium oxide from New Zealand, and inspired Marsden to extend the South Island search for radioactive minerals.

Marsden's uranium survey continued the next year, and in March 1945 the DSIR chartered the government ship *New Golden Hind*. The ship sailed down the South Island's east coast and around Bluff to investigate the eight sounds from Milford Sound to Nancy Sound. Once in the fiords, the scientists explored the coasts in outboard motor boats, testing the radioactivity of coastal rocks with their Geiger counters. One team set up camp at Gillespies Beach (downstream from Fox Glacier on the West Coast) whose black sands had shown significant levels of radioactivity. Jim McCahon, a young physicist seconded to the DSIR, later recalled the work:

[It was] very hot, so we worked most of the time nearly naked and, of course, any time a little cloud came over the sun we were free feed for the sandflies. When we had been there for about ten days, Dr Ernie Marsden of the DSIR called in to see how we were getting on. This whole project was a favourite of his. We told him of the sandfly troubles. He went away to his car and came back with a bottle labelled dimethyl phthalate — the stuff that is nowadays called 'dimp'. He had bludged this from the airforce up in the islands and he gave it to us for protection. It was an absolute godsend.[24]

THE 'ATOMIC AGE' BEGINS

By the end of May 1945, scientists and technicians at Los Alamos had received enough plutonium from the plant at Hanford to begin assembling a plutonium-based bomb. In a matter of weeks they had the makings of the world's first nuclear weapon. To make the implosion bomb, codenamed Trinity, a single sphere of plutonium was compressed by an explosive charge to create criticality. On 16 July, the bomb was tested in a remote corner of the Alamogordo Air Base in the New Mexico desert. The explosion, witnessed by busloads of Los Alamos scientists and other observers, yielded energy close to that of 20,000 tons of TNT, vapourising the steel tower holding the bomb, fusing the desert sand below it into a crater of radioactive green glass, and sending a mushroom cloud 12 kilometres into the desert sky.

The success of the test paved the way for the planned use of nuclear weapons against Japan, just as metallurgists at Los Alamos were assembling the final components in Little Boy (a uranium-based bomb) and Fat Man (a second plutonium bomb).

On 6 August, an American B-29 bomber took off from the American base at Tinian, in the Northern Mariana Islands, and dropped a 3-metre-long bomb containing 60 kilograms of uranium-235 on the city of Hiroshima. The bomb exploded about 600 metres above the ground, producing a rapidly expanding fireball that incinerated everything and everyone within a radius of about 1.6 kilometres, and spread fires across a further 11 square kilometres. The press release issued by the White House later that day described the bomb as 'the greatest achievement of organized science in history'.[25] Three days later, an even more powerful plutonium-based fission bomb was exploded over Nagasaki. Burn injuries and radiation affected many of the initial survivors, and by the end of 1945 an estimated 140,000 people had died from the Hiroshima bomb and 70,000 from the Nagasaki bomb.

Few New Zealanders would have connected the work of New Zealand scientists with the dropping of the first nuclear bombs, but an official New Zealand press release, issued on 13 August 1945, linked the bombs to Rutherford's early work, provided information about Marsden's uranium survey, and outlined the role of the New Zealand scientists working in North America, saying how New Zealanders should be proud that her scientists were at the forefront of this latest development.

Japan agreed to surrender the day after Nagasaki was bombed. The general reaction in New Zealand, and in other Allied countries, was one of jubilation and relief. The war that had taken more than 11,000 New Zealand lives and had an impact on every aspect of society was finally over. While it was marvelled at that a single bomb dropped from a great height could cause such devastation, there was initially no awareness of how fundamentally different these bombs were: the conventional bombings of cities like Tokyo, Hamburg and Dresden had produced more casualties than in Hiroshima or Nagasaki, and the longer-term effects of radiation from the bombs were not yet known.

Even people who recognised the horrific aspect of the new type of bomb were able to put a positive spin on it: the *New Zealand Listener* editorial of 17 August described the use of the atomic bomb as having 'sickened many people and given others a faint gleam of hope', but took the stance that it was 'justifiable to hope as well as to shudder' — there was hope that the atomic bomb could mean the end of war.[26] There were a few letters to the editor about the bomb — mostly expressing the hope that it could mean an end to war forever — but most New Zealanders were focused on relatives still overseas and letters focused on the practical necessities of coping with wartime shortages like how to re-waterproof an old raincoat, or make a fowl-house from old sacks and a wooden frame. Some people, however, realised the enormity of this new scientific and military development. A few days after the bombings, philosophy lecturer Karl Popper addressed a packed lecture hall at Canterbury University College with the words 'when the first atomic bomb exploded, the world as we have come to know it came, I believe, to an end'.[27]

Robin Williams was holidaying in California with his wife when they saw the news headline announcing the Hiroshima bombing, and he realised that it was the result of the project he had been working on. Williams remembers no

After making a Geiger counter in the DSIR laboratory, Jim McCahon helped search for uranium on the beaches of the West Coast of the South Island. He later joined the staff of the Dominion Physical Laboratory and became a member of the Campaign for Nuclear Disarmament. Courtesy Jim McCahon.

discussion of moral issues among the British scientists in his team, and soon after he returned to Berkeley the assembled team began to disperse. Williams was the first of the New Zealand team to leave North America, travelling to Cambridge to take a PhD in mathematics. Page, who had earlier described his secondment to the United States as 'a chance of a lifetime', transferred to Montreal to work with Watson-Munro's team.[28]

Jim McCahon, who had been employed on Marsden's South Island uranium search, was in the laboratory in Wellington, analysing samples taken in the search, when he heard a radio bulletin announcing the Hiroshima bombing. He later described himself and his colleagues as having been astounded. When the uranium survey was first announced, they had found the German paper detailing the initial discovery of uranium fission in which 'they had surmised that this could be used as a source of enormous amounts of energy but . . . not as an explosion. So we were thinking of nuclear power supplies . . . but not bombs.'[29]

A Labour Government, under Prime Minister Peter Fraser, was in power in New Zealand when Japan was bombed. There was no big discussion about the atomic bomb in Parliament, but various politicians referred to it, amid debate about other issues, in a mostly positive light. Robert

Most responses to the dropping of the atomic bombs on Hiroshima and Nagasaki were positive, particularly among those involved in the Second World War. Major Clarence Skinner, a minister in the Labour Government, shown here (left) in Cairo in 1941 with Prime Minister Peter Fraser (centre), claimed proudly that the British and American scientists didn't take long 'to show the Japanese scientists who could do the best', and he offered 'my gratitude for what has happened during the last few weeks — the ending of the war'. PA-Coll-4161, Alexander Turnbull Library, Wellington, New Zealand.

Macfarlane, Labour MP for Christchurch South, accused people who wrote letters to the newspaper expressing indignation about the use of the bomb of being 'Pacifists' — a derogatory term during wartime — and saying that apart from its use as a destructive weapon, the atomic bomb 'might have opened a new era of development for the people of the world, and so some good may arise from its invention'.[30] Major Clarence Skinner, a minister in the Labour Government, spoke proudly of the work of the British and American scientists, who didn't take long 'to show the Japanese scientists who could do the best'. He continued by saying, 'A couple of doses of atomic bomb worked the oracle, and now we see these Japanese taking orders from mere mortal men. I join with other members in offering my gratitude for what has happened during the last few weeks — the ending of the war.'[31] Another Labour MP, Edward Cullen, had a less positive view and expressed his opinion that the atomic bomb was 'a frightful instrument against humanity'.[32]

Parliament did, however, take action to control any uranium resources New Zealand might have. Under Marsden's recommendation, and following the advice of the British DSIR, New Zealand's Atomic Energy Act was passed on 7 December 1945 to give the State full ownership and control over uranium and other radioactive elements, with the Minister of Mines having power to control the mining and disposal of uranium-bearing rock and its products. The military and economic importance of uranium was now recognised around the world, and Canada, South Africa, India, Australia and the United States all introduced similar legislation at about this time.

Scientists were quick to realise the dangers of this new weapon. In September 1945, Williams and Page were among thirteen British Berkeley scientists, including Wilkins, Oliphant and Massey, who, acting on their belief that 'the advent of this new weapon of destruction ought to be the signal for renewed efforts to achieve lasting world peace' signed a letter to British Prime Minister Clement Attlee calling for international control of the use of atomic energy and urging co-operation with Russian and other scientists.[33]

This desire for international scientific co-operation with regard to nuclear weapons was widespread. That same month, Marsden advised the Minister of Scientific and Industrial Research of resolutions passed at a meeting of the New Zealand Council of Scientific and Industrial Research (an advisory council established at the same time as the DSIR) calling for the development and control of atomic energy to come under the aegis of the United Nations, with the results of research available to all people. 'Any attempt at secrecy in this epoch-making field of research is fraught with the gravest possible danger to our civilisation,' Marsden added.[34] The New Zealand scientists' attitudes were in line with a worldwide movement that included many of the scientists who had worked on the North American nuclear projects. As the New Zealand Second World War narrative on atomic energy colourfully put it in 1948, these international groupings of scientists were driven by 'fear of the immediate physical consequences of an armaments race in the atomic age, fear of the destruction of civilisation and of the best fruits of the human spirit, fear that science might be shackled by security and other regulations and enfeebled by its prostitution to nationalistic military ends'.[35]

In January 1946, less than six months after the dropping of the first atomic bombs, New Zealand was one of 51 nations represented at the first General Assembly of the United Nations. The first resolution adopted concerned the establishment of an Atomic Energy Commission, comprising the members of the Security Council, plus Canada, to deal with issues related to the peaceful uses of atomic energy and the elimination of atomic and other weapons of mass destruction. In the general debate in the plenary meeting, the New Zealand representative suggested that control of the Commission should not be left exclusively to the Security Council, as had been suggested, but should rather be the responsibility of the entire General Assembly — this way small countries like New Zealand could continue to be able to have a say on such issues — but this was not heeded. Later that year, the United States presented a proposal known as the Baruch Plan, which called for international inspection of all nuclear-related facilities to ensure they were not working on atomic weapons, and stipulated that the United States stop all weapons work and turn over its atomic energy knowledge, and existing weapons, to the United Nations.

While the DSIR scientists' work on the nuclear bomb finished once the bomb had been used, another group of New Zealanders became involved in the bomb's aftermath, with New Zealand Army engineers involved in the demolition of bomb-damaged buildings and bunkers in Hiroshima. They, along with other personnel from Jayforce — the New Zealand army brigade stationed in Japan from 1946 to 1948 as part of the British Commonwealth Occupation Force — witnessed the devastation and after-effects of the bomb. Some Jayforce personnel later recalled being told to remain inside their trucks or trains when passing close to the zone where the atomic bomb was detonated in Hiroshima, but most received no warnings about the dangers from radiation poisoning. Even troops not stationed in Hiroshima would take sightseeing trips to the devastated city, where they would 'hunt for souvenirs such as fused glass', or offer goods for sale on the local black market.[36] One officer later recalled eating oysters from a Hiroshima pearl farm.

Back in New Zealand, once the excitement of the end of the war was over, there was a growing awareness that a new age, the 'atomic age', had begun. In New Zealand, as in the rest of the Western world, the atomic age was seen as a modern and sophisticated new era. In a 1946 issue of the *New Zealand Listener*, alongside the advertisements for pointy bras, laxatives and cork-tipped cigarettes, were advertisements for Atomic Red lipstick. It seems in

The atomic age was seen as an exciting and sophisticated new era, as evidenced by Monterey's advertisements for Atomic Red lipstick. *New Zealand Listener*, 15 Feb. 1946 and 8 Mar. 1946.

appalling bad taste now to link sexuality with weapons that had killed tens of thousands of people, but the Atomic Red lipstick ads promised women they'd be 'charged with excitement . . . devastating . . . all conquering', saying women who wore the lipstick were chic and daring.

NEW ZEALAND SCIENTISTS' WORK ON NUCLEAR REACTORS

The New Zealand scientists working on the Manhattan Project had fulfilled their role and moved on, but what of the scientists in Canada? In August 1945, Norman Manssen, one of the men Marsden had recommended earlier, became available and travelled to Canada to work with the New Zealand team at Chalk River. In seeking permission to send Manssen, Marsden expressed his view that the experience he and the other scientists would gain on the project 'would pay handsome dividends to New Zealand upon their return. It will enable us to proceed upon developments of this new source of energy to our own ultimate economy.'[37]

As the Canadian nuclear energy project had developed, scientists had made a case for producing a 'zero-energy pile' — a pile that produces a chain reaction but almost no extra heat or energy. Such a pile would be

Opposite: The zero-energy experimental pile (ZEEP) at Chalk River, Canada, showing the top of the reactor's fuel rods. The fuel rods, which consisted of aluminium tubes containing natural uranium slugs, hung vertically in a 3-metre high tank of heavy water. Courtesy National Research Council Canada.

The New Zealand team designed ZEEP's control panel with its array of radiation detectors and dials and switches to monitor the reactor and control its operation. Courtesy National Research Council Canada.

simple to produce, flexible in its applications, and would have great research value, producing neutrons and radiations for use in a range of experiments. It would also provide information that could be applied to the planned larger pilot plant to produce energy. The plan was approved for a zero-energy experimental pile, or ZEEP, in which rods of aluminium-sheathed uranium would be placed in a tank of heavy water surrounded by a graphite reflector. Watson-Munro, who was leading the small New Zealand team at Chalk River, was second-in-command of the team of six scientists and engineers assigned to the detailed engineering of ZEEP. By September 1945, ZEEP, the first nuclear reactor built outside the United States, was complete, and Watson-Munro, with the help of Allan and Fergusson, had played a major role in its construction, particularly by designing the electronic control equipment for the reactor.

New Zealanders' involvement in the development of nuclear energy continued. After the war, the British Government turned their focus to a United Kingdom-based atomic energy research project, and in 1946 Watson-Munro and three of the other New Zealanders — Fergusson, Page and Walker — left Canada for the newly established United Kingdom Atomic

Energy Research Establishment in Harwell. The remaining New Zealanders — George, Young, Allan and Manssen — initially remained in Canada to continue work on the pilot pile, which became known as the NRX (National Research eXperimental) reactor. While Young was put in charge of the Montreal workshops, George worked on designing nuclear instruments and equipment, and Manssen supervised the installation and commissioning of the electronic and electrical equipment for the NRX reactor, which went critical in July 1947 and for many years was considered one of the best research reactors in the world.

In the United Kingdom, Watson-Munro took charge of the construction of a graphite low-energy experimental pile, or GLEEP, the first nuclear reactor in the United Kingdom, which was completed in August 1947 and was intended for use for experimental purposes as well as for the production of radioactive isotopes. Page, also working at Harwell, focused on design and layout of the reactor's instrumentation and control equipment. Allan and Manssen transferred to Harwell after completing their work at Chalk River.

AN ATOMIC PILE FOR NEW ZEALAND?

After the war, Marsden started to enact his vision for a nuclear New Zealand. If he could not be part of the big science taking place in Europe and America, he would make it happen at home. In January 1946, Marsden gained Cabinet approval to establish a new team of ten nuclear scientists at the Dominion Physical Laboratory. Their mission was to carry out fundamental and applied atomic research and advise on atomic energy and the application of isotope techniques to problems in agriculture, health and industry. The same Cabinet decision allowed for the secondment of physicists, chemists or engineers to nuclear organisations in the United Kingdom and Canada to ensure New Zealand kept up to date with new developments and techniques. Under the secondment arrangement, New Zealand would pay the officers' salaries, and in return would receive supplies of artificially radioactive elements, technical information, and co-operation and liaison work in connection with the laboratory in New Zealand. An annual budget of £19,000 was allocated to implement these proposals.

Even before this approval was granted, Marsden had organised a second *New Golden Hind* expedition to complete the initial uranium survey. This expedition was not secret, and Bob Semple and Daniel Sullivan, ministers in Peter Fraser's Labour Government, posed for photographs with the scientists before the ship left Wellington. Dick Willett, of the New Zealand Geological Survey, led the January 1946 expedition that searched the rocks, beaches and gravels in the southern sounds from Preservation Inlet up to Thompson Sound. He took geologist Harold Wellman, and three geophysicists, Jim McCahon, Kemp Fowler and Graham Fraser, to operate the Geiger counters and measure the radioactivity of the rocks and dredgings. The bearded scientists and Rarotongan crew supplemented their diet of bully beef, potatoes and pumpkin with fresh fish, crayfish and swan: 'Shooting at [swans] from a lurching dinghy is useless; the method is to run down a young one unable to fly and brain it with an oar,' said Fowler. [38]

Over the course of the two-year uranium survey, the geologists tested the radioactivity of rocks and beaches along nearly 1600 kilometres of coast; the sands and gravels from more than 100 streams and rivers; and the concentrates from more than 20 sluicing and gold-dredging claims in Nelson, Westland, Otago and Southland. As the only result of the survey, uranium-bearing minerals were found in gold-dredge tailings on the West Coast, but they were not there in high enough concentrations to make it economically worthwhile to extract them.

Without a domestic source of uranium, was there still hope for a nuclear reactor for New Zealand? New Zealand might not have had any promising uranium deposits, but it did have a skilled group of scientists, and — as Marsden had pointed out in a September 1945 letter to Edward Appleton, the director of the British Tube Alloys Project — 'they cannot unlearn the things they picked up from their work with you and cannot well be prevented from using the knowledge . . .'.[39]

Marsden had made a successful offer for New Zealand scientists to continue to be seconded to British nuclear projects, in return for 'reasonable information received and the possible supply in the next year or two of labelled atoms for experimental work in medical, agricultural and chemical fields'.[40] As he had said, the New Zealand scientists could not 'unlearn' what they now knew about nuclear energy: following his work on GLEEP and ZEEP, Watson-Munro was now one of the world's experts on construction of nuclear piles, and despite the British–American agreement not to share nuclear technology with other countries it was going to be impossible to keep this technology from New Zealand. Cockcroft was aware of this when

The *New Golden Hind* sets off from Wellington on its second uranium survey, in January 1946. The men standing on the left are Minister of Works Bob Semple, and the Minister in Charge of the DSIR Daniel Sullivan, who were happy to pose with the scientists charged with searching for uranium in New Zealand. Seated in the front are physicist Jim McCahon, geologist Dick Willett, and Harry Allan, Director of the DSIR's Botany Division. In the back are a ship's officer, A. S. C. Wright, Captain Cole, the radio operator, physicist Kemp Fowler, a ministerial secretary, geologist Harold Wellman, Director of the Geological Survey Mont Ongley, and physicist Graham Fraser. Courtesy Simon Nathan.

he wrote to Marsden in May 1947. After voicing appreciation of the New Zealanders who had come to the United Kingdom to work on GLEEP, Cockcroft expressed his hope that New Zealand would 'build a Gleep, or perhaps something a little more powerful, in your area of the world. Munro would be well able to do this for you, and I should think it would be possible to arrange for the basic materials to be provided.'[41] Cockcroft at this time, as part of a policy of an Empire-wide approach to defence science, was also open to the idea of a Commonwealth nuclear reactor to produce plutonium for British bombs.

Before returning to New Zealand, Watson-Munro, in consultation with Marsden, submitted a report to the New Zealand Government on the construction of a low-energy atomic pile in New Zealand. The pile was recommended on two grounds: for the production of radioisotopes for industrial and agricultural research; and to serve as the nucleus of an atomic research project. Marsden also believed the pile would provide a 'long term contribution to Commonwealth defence'.[42] In August 1947, based on Marsden and Watson-Munro's report, New Zealand's newly-established Atomic Energy Research Committee recommended the construction of an Australasian low-energy pile in New Zealand. The pile was proposed to be used 'for the provision of certain radio isotopes whose life was short and which are needed for medical and research purposes, and to afford, as a general strategic reserve, skill and experience to New Zealand scientists in atomic energy development'.[43]

Australia had already begun its own research into nuclear energy and the availability of uranium reserves for mining. While Australia was more immediately enthusiastic than New Zealand about embracing nuclear technology — they asked the United Kingdom directly for information on using atomic energy — they did not have the group of scientists experienced in working on nuclear reactors that New Zealand now had. Mark Oliphant and a small group of Australians had worked at Berkeley on the electromagnetic separation of uranium for the Manhattan Project, but only one Australian engineer had worked on the Canadian nuclear energy project. After helping to set up the British Atomic Energy Research Establishment at Harwell, Oliphant returned to Australia to head a school of nuclear physics at the new Australian National University in Canberra.

The American post-war secrecy led the Commonwealth countries — including New Zealand, Australia and the United Kingdom — to side together, often conducting research on projects without American knowledge. In a way, this encouraged the sense that there was a choice to be made between the United Kingdom and the United States; New Zealand's scientific loyalty was clearly with the United Kingdom.

In September 1947, Marsden left his position as secretary of the DSIR to become the New Zealand DSIR's scientific advisor in London. It seems an unlikely move for someone of Marsden's seniority to move from a role as head of department to a liaison role, but his letters over this time reveal some motives. It seems that Marsden, to some extent, felt unappreciated by those above him in the Public Service Commission. He also strongly supported Commonwealth co-operation with regard to defence science, and felt he could play a stronger role in cementing co-operation between New Zealand and the United Kingdom if he were in London. In his personal letters, he mentioned several British requests for a more senior officer to be sent from New Zealand as a scientific representative. On top of this was Marsden's already demonstrated desire to be involved in the big science that was happening in the United Kingdom, and, with New Zealand so far having an ambivalent attitude to proceeding with plans for a nuclear reactor, Marsden perhaps felt he could push things along further from the United Kingdom.

When Marsden arrived in London, he and Watson-Munro met Lord Portal, head of the Atomic Energy (Review of Production) Committee, to talk about the Commonwealth atomic pile. They discussed the advantages of a small atomic pile in New Zealand for research purposes, to be followed by a large power production pile in Australia, 'capable of producing fissile materials suitable for the manufacture of atomic bombs'.[44] At this time, the Atomic Energy Production Organisation was erecting three large atomic piles in the north of England, all capable of producing fissile material for military purposes. It was suggested that it might be strategic to locate one of these in Australia.

Marsden's enthusiasm for New Zealand's assistance to British nuclear defence plans can be seen as incongruous when measured against his prior support for the 1946 Baruch Plan. In a 1947 speech, Marsden, who advocated atomic energy as being of 'untold benefit to the world', said that it was not, however, safe to develop atomic energy on a worldwide scale until there was a practical and enforceable agreement that it would not be used for atomic bombs.[45] No such agreement was put in place, and his stated views on atomic weapons seem to conflict with his concurrent plans for development of a nuclear reactor in New Zealand, which he promoted as being of defence significance to the Commonwealth. A reading of his letters, however, suggests that his approach was a result of a compromise in his beliefs. At the start of the Cold War, the United States and the Soviet Union were proceeding with development of nuclear weapons, and Marsden believed that a third nuclear power — the United Kingdom (with the support of the Commonwealth) — would make for a more balanced and stable world, with 'more safety in three strong world groups than two'.[46]

On receiving sympathetic responses to the proposal for a New Zealand atomic pile from both Lord Portal and John Cockcroft, who was now director of the Atomic Energy Research Establishment at Harwell, Marsden was tremendously excited. He admitted he had initially thought the reactor proposal was an 'ambitious dream', but was now convinced it would be 'a statesmanlike step to take at higher levels with enormous repercussions for the good of our country'. Marsden, writing from London, believed the reason New Zealand was getting more support than Australia was 'partly sentimental, because of its origination by Rutherford, partly practical because of the record of our boys here'.[47]

In response to a ministerial request, Marsden and Watson-Munro provided an advisory report, which was agreed to by Cockcroft, on the construction and use of an atomic pile in New Zealand. The November 1947 report recommended a graphite uranium pile, costing £100,000 to construct and up to £35,000 a year to run. The project would use the skills of the New Zealand scientists who had worked on the North American nuclear programmes and would take one to two years to build. They made it clear that the proposed pile 'would not produce an atomic bomb' but 'would contribute substantially in the future to the security of the Commonwealth'.[48] The Minister of Scientific and Industrial Research, however, was critical of the report, questioning the need for a New Zealand pile on the basis that radioisotopes were already available from overseas and that New Zealand scientists would be best trained in more sophisticated offshore facilities.

Marsden and Watson-Munro were not deterred, however, and in March the following year Watson-Munro expanded on this report, again outlining the issue of many isotopes being too short-lived to import from overseas, and emphasising the benefits of a pile as a practice run for a future atomic energy project. At this time, New Zealand was having great difficulty securing radioisotopes from overseas because of the very short half-life of some of them, and the large amount of shielding required for some overseas shipments (a 1-pound shipment of radioactive cobalt brought to New Zealand on a Navy frigate was reported as requiring three-quarters of a ton of concrete shielding). The defence applications were also raised, partly from having a centre of nuclear physics away from the northern hemisphere, but also in giving New Zealand scientists expertise to help deal with 'an atomic dust attack on a major New Zealand city or water supply'.[49] But this time, Henry Tizard, scientific advisor to the British Ministry of Defence, gave the proposal a lukewarm reception, telling the ever-enthusiastic Marsden that the defence arguments in favour of the pile were weak.

Prime Minister Peter Fraser then sought the opinion of the British prime minister on the value of the project. Clement Attlee replied favourably, saying the project would be of 'advantage to the Commonwealth' and offering the assistance of the United Kingdom Government.[50] Fraser travelled to the United Kingdom in December 1948, and visited the low-energy atomic pile at Harwell. Marsden briefed Fraser on his visit to Harwell, where he would see the GLEEP in action, and meet the five New Zealanders now working there. Marsden also briefed Fraser on the benefits of a low-energy atomic pile, pointing out that, while South Africa and Australia could more easily obtain radioisotopes from the United Kingdom, in New Zealand's case a change of aeroplane was involved, resulting in more difficult organisation and the loss of valuable time, having a great impact on the important short-lived isotopes of iodine and phosphorus. Marsden added that the pile suggested for New Zealand 'is about 10 times as powerful as that of GLEEP . . . but will be less bulky, and will involve no danger whatsoever to anyone living near it'.[51] Marsden suggested somewhere near Lincoln College in Canterbury as a suitable site, as it was close to the airport and medical school and would be a good place from which to distribute to the rest of the country.

Marsden continued to advocate for construction of an atomic pile in New Zealand. But given his absence from New Zealand, and — despite Attlee's offer of assistance — with limited government support for an atomic pile, many of the DSIR's original nuclear sciences team moved into other areas of research. Charles Watson-Munro, George Page and another New Zealand physicist who had worked at Harwell, Cliff Dalton, eventually moved to the Australian Atomic Energy Commission Research Establishment, where Watson-Munro became chief scientist. The DSIR nuclear sciences team Marsden had established continued, although rather than operating a research reactor they focused on measuring environmental radioactivity, using radioactive tracers, and experimenting with radiocarbon dating.

The atomic bomb — with some thanks to Rutherford's earlier genius and the assistance of a small group of New Zealand scientists — had won the Second World War, and New Zealanders were grateful in that regard. But in the years that followed, information exchange came to a halt, with a breakdown in British–American scientific relations, the start of an international arms race, and the Cold War between the United States and the Soviet Union. Information about atomic energy and atomic weapons was top-secret as first the Soviet Union, then the United Kingdom, then France and China, began their own development of nuclear weapons.

How would New Zealanders react when the United States, and then the United Kingdom, started testing new atomic weapons in the Pacific, New Zealand's backyard?

3

★

Cold War and
red-hot science
The nuclear age
comes to the Pacific

*As it detonated, I felt the heat and saw the flash through my eyelids and
fingers. About half a minute later we were told to face the burst. All cloud
cover had gone, and the coloured fireball was climbing rapidly upwards. It was
an impressive sight . . .* — MAURICE HAYMAN, OPERATION GRAPPLE, 1958[1]

*. . . The inhabitants might be advised to eat only coconuts during the
period of the tests.* — DEPARTMENT OF HEALTH RECOMMENDATION, 1957[2]

The 1950s are often remembered as a golden age of prosperity and
optimism, with full employment for men, a baby boom and a rise
in consumer culture. But it was also the decade when the Cold
War brought American and British bomb tests into New Zealand's Pacific
backyard and phrases like 'thermonuclear war' and 'radioactive fallout'
entered New Zealanders' vocabulary. While the end of the Second World War
was a cause for optimism, the Cold War became a new source of fear. Most
New Zealanders, though, saw the perceived threat from the Soviet Union
and other Communist nations as justification for the emerging nuclear arms
race, even when the search for new bomb-testing territories led the Pacific to
become what Stewart Firth has called a 'nuclear playground'.[3]

The first two nations to test nuclear weapons in the Pacific were the United States and the United Kingdom. Historian David McIntyre has described New Zealand's post-war period as one of 'dual dependency' on the United Kingdom and on the United States, New Zealand's powerful new friend and protector.[4] When it came to their Pacific nuclear-testing programmes, however, it was clear which country was considered most favourably: the Government offered logistical support to the British nuclear-testing programme, and New Zealanders took pride in British scientific and military achievements. In contrast, the United States nuclear-testing programme was acknowledged, rather than celebrated, as being necessary to their maintaining supremacy in the arms race. In both relationships, however, New Zealand showed strong signs of independence, saying no to specific American and British requests if there was no benefit to New Zealand, even when it challenged traditional or emerging loyalties.

THE PACIFIC 'NUCLEAR PLAYGROUND'

The British–American wartime co-operation on nuclear weapons and nuclear energy development did not last. The United States Atomic Energy Act of 1946 transferred control of atomic research from a military to a civilian organisation, the United States Atomic Energy Commission (USAEC). Following reports of Soviet infiltration into the Canadian nuclear energy project, the Act prohibited the transfer of nuclear technology to other countries, ending the British–American partnership agreed to in 1943. The United Kingdom had already decided to move its atomic energy research and development programme from Chalk River, in Canada, to Harwell, in Oxfordshire, and now made plans for a British weapons development laboratory as well.

Post-war New Zealand was changing politically, and the country was beginning to assert greater independence. Known as a Dominion of the British Empire from 1907, New Zealand now dropped 'Dominion of' and was known solely as 'New Zealand'. With the passing of the Statute of Westminster Adoption Act in 1947, New Zealand Parliament lost its legal subordinance to the Parliament of the United Kingdom and gained control of its own rules of government.

While the relationship with the United Kingdom was still strong — it was the destination of two-thirds of New Zealand's exports and the source of

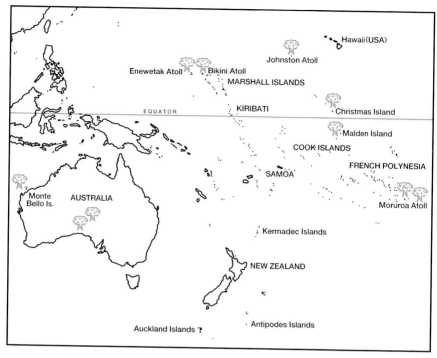

In the late 1940s and the 1950s the Pacific became what Stewart Firth has called a 'nuclear playground' for the nuclear powers. This map shows sites of major Pacific nuclear bomb tests by the United Kingdom, the United States and France.
Map after Stewart Firth, *Nuclear Playground*, Allen & Unwin, Sydney, 1987.

nearly half of New Zealand's permanent immigrants — the Pacific War with Japan had highlighted New Zealand's need for the United States' protection. New Zealand had opened a legation in Washington in 1941, but the relationship strengthened after 1951, when the ANZUS Treaty was signed. This security agreement between New Zealand, Australia and the United States offered the United States support for their 'soft peace' with Japan, while giving Australia and New Zealand a guarantee of protection against any Japanese military threat. There was also growing anxiety, particularly from Australia and the United States, about Communist expansion into South East Asia: the People's Republic of China had been formed in 1949 after a long civil war, making the most populous country in the world now Communist.

Despite the importance of the relationship with the United States in protecting New Zealand from Japan or the perceived threat of Communism, when the United Kingdom and the United States began testing nuclear bombs in the Pacific it was clear that New Zealand's support lay primarily with the United Kingdom.

AMERICAN PACIFIC NUCLEAR TESTS BEGIN

According to Cold War historian John Lewis Gaddis, victory in the Second World War brought no sense of security to the victors. After the successful deployment of nuclear weapons in the war against Japan, the United States knew that other countries would be trying to develop the new technology and wanted to stay one step ahead of them. The United States continued to develop nuclear bombs and test them in Nevada, and in new Pacific testing grounds in the Marshall Islands.

On 1 July 1946, the United States exploded a nuclear bomb in the atmosphere above a fleet of captured and obsolete warships in Bikini Atoll to determine the effect of an atomic burst on Navy ships and personnel. The target fleet of 70 vessels carried thousands of animals — goats, pigs, rats, mice and guinea pigs — so scientists could observe the effects of the shockwave and visible, thermal and nuclear radiation. The explosion created an enormous fireball that rapidly mushroomed skywards. Most newspaper reports were matter-of-fact about the details of the first Bikini Atoll bomb test, but a week later a *New Zealand Listener* editorial described the test as 'deeply depressing' — the birth of the atomic bomb had not led to the end of war, but to the prospect of more horrible war.[5] The bomb tests were not discussed in New Zealand's Parliament, although Thomas Bloodworth, an Auckland politician and a member of New Zealand's Legislative Council, described the atomic bomb as putting 'fear and apprehension into the mind of every thinking man and woman'.[6]

This first Bikini Atoll test was followed by a second a few weeks later. New testing sites were found at Enewetak Atoll, a Marshall Islands coral atoll 350 kilometres west of Bikini, and at Johnston Atoll, an uninhabited American territory in the North Pacific, and between 1948 and 1958, after which there was a temporary moratorium on nuclear testing, the United States tested nearly 70 nuclear weapons. Meanwhile, the Soviet Union had its own nuclear weapons programme, and with the explosion of a plutonium fission bomb in the Kazakhstan desert in August 1949 — news of which was received in New Zealand as an inevitability — the nuclear arms race began.

AMERICAN ATOMIC WEAPON TESTS IN THE PACIFIC, 1946–62

OPERATION AND LOCATION (LOCAL DATE)	ESTIMATED YIELD
Bikini Atoll — Operation Crossroads 1 July 1946 25 July 1946	 21 kt 21 kt
Enewetak Atoll — Operation Sandstone 15 April 1948 1 May 1948 15 May 1948	 37 kt 49 kt 18 kt
Enewetak Atoll — Operation Greenhouse 8 April 1951 21 April 1951 9 May 1951 25 May 1951	 81 kt 47 kt 225 kt 45.5 kt
Enewetak Atoll — Operation Ivy 1 November 1952 16 November 1952	 10.4 Mt 500 kt
Bikini and Enewetak atolls — Operation Castle 1 March 1954 27 March 1954 7 April 1954 26 April 1954 5 May 1954 14 May 1954	 15 Mt 11 Mt 110 kt 6.9 Mt 13.5 Mt 1.7 Mt
Bikini and Enewetak atolls — Operation Redwing 17 weapons tested 21 May and 22 July 1956	 0.19 kt–5 Mt
Bikini, Enewetak and Johnston atolls — Operation Hardtack 36 weapons tested 28 April and 18 August 1958	 0.02 kt–9.3 Mt
Christmas Island and Johnston Atoll — Operation Dominic 36 weapons tested 26 April and 4 November 1962	 40 kt–8.3 Mt

Note: kt = kiloton and Mt = megaton: units of explosive force equivalent to 1000 and 1 million tons of TNT, respectively.

Source: Adapted from S. L. Simon and W. L. Robison, 'A compilation of nuclear weapons test detonation data for U.S. Pacific Ocean tests', *Journal of Health Physics*, 73(1), 1997, pp. 258–64. Downloaded 26 Mar. 2009 from www.hss.energy.gov/HealthSafety/IHS/marshall/marsh/journal/rpt-22.pdf.

BRITISH NUCLEAR WEAPONS DEVELOPMENT

In the United Kingdom, scientists and engineers at the Atomic Energy Research Establishment at Harwell continued to research all aspects of atomic energy. New plants were established to produce plutonium and uranium-235, primarily for military purposes. Following Britain's 1947 decision to proceed with developing their own nuclear weapons, the Atomic Weapons Research Establishment was created at Aldermaston, under the leadership of William Penney.

On the other side of the world, New Zealand was assisting Britain's nuclear programme in a project designed to use Wairakei's geothermal steam in a plant producing heavy water. Heavy water was required primarily as a moderator — to dampen and control the motions of the high-energy neutrons released from the fuel source — for use in Britain's research reactors. The nucleus of heavy hydrogen, or deuterium, was the optimum size to slow fast neutrons without absorbing them, making heavy water better than regular water for controlling nuclear reactions. Because of its slightly higher boiling point — heavy water boils at 101°C — heavy water could be separated out from ordinary water, where it occurs in concentrations of about 136 parts per million, by fractional distillation, a process that requires large quantities of heat. The attraction of using geothermal energy was that Nature had provided vast quantities of hot water; in other locations, a hydroelectric or coal-burning power plant would be needed to heat the water.

The Wairakei project had first emerged in 1945 when, following suggestions made by local authorities, the Government had asked the DSIR to investigate the commercial development of the geothermal resources in the Rotorua area. One of the many propositions investigated was the production of heavy water. Following investigations by a small DSIR team (which included Jim McCahon of the Dominion Physical Laboratory, who had also worked on the wartime uranium survey), Ernest Marsden informed the United Kingdom of the possible availability of large quantities of heat for the production of heavy water, and was encouraged to proceed with the studies. By 1947, after two dry years that led to electricity shortages, geothermal power was also emerging as an attractive option for electricity generation. The Wairakei field, next to the Waikato River north of Taupo, was considered the most promising of New Zealand's geothermal fields.

Marsden, now working as scientific liaison officer in London, supported the project. In 1949 he reiterated his earlier suggestion to John Cockcroft — who was now director of the British Atomic Energy Research Establishment — that New Zealand's geothermal steam could be used to concentrate heavy water through fractional distillation. Cockcroft was receptive to Marsden's suggestion, and a DSIR scientist, Tony McWilliams, travelled to Harwell to study the distillation of the heavy-water fraction from naturally occurring water through use of geothermal steam.

In March 1952, the New Zealand Government received formal advice that the British authorities attached great importance to the development of additional supplies of heavy water and requested an urgent and thorough survey of its potential production in New Zealand. Later that year, a scientific drilling project calculated that the Wairakei field had enough steam to generate 20 megawatts of electricity, which was not a huge amount of power but was comparable to the output of some of the country's smaller hydroelectric power stations.*

Marsden continued to encourage the project, liaising between Harwell, the DSIR and the New Zealand Prime Minister's Department. McWilliams experimented with fractional distillation at Harwell, and in New Zealand the DSIR investigated the geothermal field and the properties of the fluid it produced.

Britain had its own nuclear bombs ready to test by 1952. But with no suitable domestic space for testing them, Britain turned to the Commonwealth and decided on the uninhabited Monte Bello Islands off Australia's north-west coast. Australian Prime Minister Robert Menzies readily agreed to Clement Attlee's request to use the islands: an Australian Government press release later summed up Australia's attitudes to British nuclear testing by saying 'England has the bomb and the know-how; we have the open spaces, much technical skill and great willingness to help the Motherland'.[7]

Wayne Reynolds argues that in the post-war period, as Britain developed nuclear weapons without support from the United States, the Empire — as a source of raw materials, manpower and testing zones — became vital to British survival.[8] Australia was happy to offer support: with its vast deposits of uranium it had plans to develop its own nuclear weapons, and so support for the British nuclear programme had a strong element of self-interest. Because of its proximity to Asia, Australia also had more fear of Chinese Communism than New Zealand did, and therefore more of a willingness to come under the protection of the 'nuclear umbrella'. New Zealand's much lower level of support for the British nuclear-testing programme can on the surface be seen to be a result of having less to offer — New Zealand did not have Australia's supposedly 'empty' spaces or uranium deposits — but nor

---

* When the Wairakei Power Station came online in 1958, it produced 145 megawatts of power.

did New Zealand have ambitions for its own nuclear weapons. New Zealand politicians, therefore, said yes to British requests when it was politically or economically expedient to do so, and no when it was not.

New Zealand's military assistance to the British nuclear programme should also be seen in the context of other Cold War military co-operation. New Zealand had already played its part in operations and wars to resist the perceived Communist threat; for example, by assisting Britain with the Berlin airlift in 1948, and fighting alongside British and Australian soldiers in the Korean War from 1950. It would, therefore, have been surprising if New Zealand had not offered some level of assistance with a British military programme so close to home.

BRITISH ATOMIC WEAPON TESTS IN AUSTRALIA, 1952–57

OPERATION AND LOCATION (LOCAL DATE)	ESTIMATED YIELD
Monte Bello Islands, WA — Operation Hurricane 3 October 1952	25 kt
Emu Field, SA — Operation Totem 15 October 1953 27 October 1953	10 kt 8 kt
Monte Bello Islands, WA — Operation Mosaic 16 May 1956 16 June 1956	15 kt 60 kt
Maralinga, SA — Operation Buffalo 27 September 1956 4 October 1956 11 October 1956 22 October 1956	15 kt 1.5 kt 3 kt 10 kt
Maralinga, SA — Operation Antler 14 September 1957 25 September 1957 9 October 1957	1 kt 6 kt 25 kt

Note: kt = kiloton and Mt = megaton: units of explosive force equivalent to 1000 and 1 million tons of TNT, respectively.

Source: Adapted from: Lorna Arnold, *A Very Special Relationship: British Atomic Weapon Trials in Australia*, HMSO, London, 1987; and Roger Cross, *Fallout: Hedley Marston and the British Bomb Tests in Australia*, Wakefield Press, Kent Town, 2001.

The United Kingdom's first nuclear weapon detonated on 3 October 1952, in the Monte Bello Islands. While the United States' 1946 Bikini Atoll tests had been described by the *New Zealand Listener* as 'depressing', and news of the 1949 Soviet test received as an inevitability, New Zealand's response to the first British test was positive. 'Unless the United States has something up her sleeve Britain has taken the world lead in the race for tactical atomic weapons,' said *The Dominion* proudly, citing the 'unusually large' area of ground-level destruction caused by the blast.[9] The newspaper's editorial noted that 'British people everywhere [and in the 1950s that included New Zealanders] will doff their hats to the quiet, clever "back-room boys" who made this achievement possible'.[10] The United States did, of course, have something 'up her sleeve' — they were already far ahead of the British programme and were about to start testing a hydrogen bomb.

The next British test series, codenamed Totem, was at Emu Field, a desert location 500 kilometres north-west of Woomera, in South Australia. After the first shot of the Totem series, on 15 October 1953, the wind blew a narrow radioactive plume to the north-east. Despite fallout — radioactive particles generated in the nuclear explosion — being detected in New Zealand, there was a romantic tone to the local newspaper reports of the blast. An NZPA report published in *The Press* read:

> As the weapon exploded, the whole countryside was deluged with light, dimming the early morning sun. The fireball was lit for a time by vivid internal flashes. It turned into a column of peach-coloured smoke, which soon took on the conventional mushroom shape.[11]

The nuclear tests were of great interest to New Zealand scientists and scientific organisations, who made their own attempts to detect radioactive fallout. Following the 1952 explosion, four Royal New Zealand Air Force (RNZAF) aircraft were dispatched from Whenuapai to collect air samples at a distance of 5600 kilometres from the Monte Bello Islands explosion; three of the four aircraft collected significant radioactivity. Under the tutelage of Charles Watson-Munro, now professor of physics at Victoria University College in Wellington, two graduate students conducted research into atmospheric radioactivity, to determine baseline levels of radon and thoron in the atmosphere and the effects of meteorological conditions on atmospheric radioactivity. Watson-Munro's own research determined that fallout from the 15 October 1953 explosion was transported rapidly to New Zealand on high-velocity, high-altitude winds, arriving in Wellington from the test zone the day after the explosion.

Staff of the Dominion X-Ray and Radium Laboratory* in Christchurch also built equipment to detect radioactivity from the British tests in Australia. Most of the particles collected after the 1953 explosions had a half-life of the order of 30 minutes, indicating that they were decay products of radon. This showed that the dangers of radioactive fallout from the explosion were not posed exclusively by products of the explosion — larger radioactive elements produced by the splitting of uranium-235 atoms — but also by radioactivity of their decay products, and *their* decay products in turn.

After another test series at the Monte Bello Islands, the United Kingdom selected a new permanent inland test site at Maralinga, in South Australia, and prepared for Operation Buffalo, which took place in September and October 1956. As well as the Australian task force and scientific staff, observers of the first Maralinga test included an 'Indoctrinee Force' of some 250 officers, including five from New Zealand, whose role was to observe the test and its aftermath, and pass on their knowledge and experiences to other members of the Armed Forces. The New Zealand military authorities were advised that the officers taking part would be 'subject to radiation hazard'. The Indoctrinee Force was stationed about 8.2 kilometres from the first blast, with their backs to the explosion. Following instructions, they turned around two seconds after the blast, and saw a massive fireball and the emerging mushroom cloud. After the blast, the indoctrinees, dressed in protective clothing, toured the blast area, examining the effect of the blast on targeted items in the area. One of the New Zealand members of the Indoctrinee Force, Lieutenant Colonel John Burns, later recalled standing in full combat gear with his back turned when a bomb was detonated. 'The heat . . . was just like the blast from opening the door on a hot oven roast. And the flash was blinding white,' he said.[12]

A report issued by the National Archives of Australia in 2001 confirmed that, along with 70 Australians, two of the New Zealand officers, including Burns, were intentionally exposed to radiation. A few days after the first Maralinga test, a group of indoctrinees entered the fallout zone as part of a project to discover which types of clothing would best protect against radioactive contamination. Burns recalled being sent to within half a mile of where the bomb was detonated, where, in enclosed rubberised suits, he and

---

* In 1947 the Travis Radiophysical Laboratory changed its name to the Dominion X-Ray and Radium Laboratory.

his fellow indoctrinees were 'marching and crawling and there was a truck passing every so often that would shower us with a bit of dust and dirt to make sure we got some of the fallout'.[13] This seems a barbaric and dangerous experiment from a modern perspective, as the soldiers' safety was in no way guaranteed. Fortunately, however, film badges worn by the New Zealand soldiers showed that each man received a one-off radiation dose no more than that received each year from natural background radioactivity.

THE BIRTH OF THE HYDROGEN BOMB

The Cold War led to an ongoing quest for more powerful bombs. A new type of nuclear weapon, which became known as the hydrogen bomb, was independently developed and tested first by the United States in 1952, then by the Soviet Union in 1953 and Britain in 1957.

There was a New Zealand connection from the start. In 1934, Ernest Rutherford and two of his Cavendish Laboratory colleagues, Mark Oliphant and Paul Harteck, had discovered the hydrogen fusion reaction, the basic principle behind the hydrogen bomb. In a letter published in *Nature*, they described how when they bombarded concentrated heavy water with accelerated deuterium nuclei, the accelerated deuterium nuclei fused with a nucleus of deuterium in the heavy water and formed a helium nucleus. The reaction, they said, released neutrons, heat and an intense burst of gamma radiation.

While Rutherford's Cavendish experiment had used more energy than it had produced, scientists determined that under extremely high temperatures unlimited quantities of heavy hydrogen could fuse into helium, releasing energy in the process. Unlike the fission reaction that created the first nuclear bombs, fusion was an extremely efficient process, and the energy output of a fusion bomb could greatly exceed that of a fission bomb.

The United States decided to proceed with the development of a hydrogen bomb in 1950, despite considerable scientific opposition to the project on the grounds that the large potential yield of such a bomb would render it a 'genocidal weapon', capable of wiping out entire cities and contaminating large areas with vast quantities of radioactivity.[14] With an arms race going on, this did not hinder the project and production continued. The first prototype hydrogen bombs, which used a fission explosion to ignite a fusion reaction, were developed at Los Alamos and tested at Enewetak Atoll in May 1951. A more advanced design — which resulted in extremely high levels of radioactive fission products being blasted into the stratosphere, from where they settled around the globe — was first tested in November 1952, resulting in a 10.4-megaton explosion. The United States conducted another test series at Bikini and Enewetak atolls in 1954. The largest detonation was a 15-megaton explosion on 1 March, which caused unexpected levels of fallout and spread harmful radiation over a large area, leading to the evacuation of American personnel and native islanders from Rongerik Atoll (an atoll 200 kilometres east of Bikini, to which Bikini's indigenous people had been moved in 1946) and other islands. The crew of a Japanese fishing boat stationed 137 kilometres away from the blast fell victim to an ash-like shower of fallout, which caused the crew to become sick from radiation poisoning; one of the crew subsequently died.

COLD WAR AND RED-HOT SCIENCE

In response to the Governor-General's speech to the opening of the New Zealand Parliament in June 1954, Warren Freer, the Labour MP for Mt Albert, noted that one of the things omitted from the speech, and also not mentioned by Prime Minister Sidney Holland, leader of New Zealand's first National Government, was the hydrogen bomb. 'Nothing has more disturbed the minds of the average men and women in the English-speaking world,' said Freer, who also referred to an earlier statement by Holland that the hydrogen bomb was the greatest promise for peace. Freer argued that Holland was wrong, and that the stance of the church leaders was correct. (That Easter, the leaders of the Anglican and Catholic churches had appealed to parliamentarians the world over to 'exercise some control over this weapon and to bring the world back to a Christian approach to international problems rather than one of cold war or, in this case, red-hot science'.) Freer called for Holland to let the New Zealand public know where he stood on the question of 'future tests of atomic weapons and hydrogen bombs, or anything more hideous'.[15]

Internationally, hydrogen bombs led people to speculate that a war waged using these weapons could mean the end of civilisation — not only could entire cities be wiped out, but the radiation could make the planet uninhabitable. In New Zealand, concern about the effect of future American tests led Cabinet to set up a scientific committee to report on the possibility of radioactive contamination of New Zealand and its Pacific island territories. The committee — which included Charles Watson-Munro, Bill Hamilton (secretary of the DSIR), and Miles Barnett (director of the Meteorological Service), as well as three senior representatives of the armed services — reported that there was no danger to New Zealand or its territories. Holland issued a reassuring press release, prepared by the committee, about the likely impact of the American bomb tests, saying that any deposits would be of 'no significance' and would pose 'no consequential risk to the inhabitants of New Zealand or its island territories'.[16]

Privately, Holland remained uneasy about the tests, although whether his concern was primarily about safety or about public opinion is hard to tell. Following the United States hydrogen bomb detonations in March 1954, Holland sent personal messages to the British and Australian prime ministers, Winston Churchill and Robert Menzies, expressing his concern about the United States tests and the risk of injury or danger to Pacific people and their food sources, and querying as to whether the United States could, in future, be asked to provide warning of such tests. In return, Churchill warned against 'any action which might impede American progress', citing their strength in nuclear weapons 'as the greatest possible deterrent against the outbreak of a third world war'.[17] With the United States nuclear weapons programme now so far ahead of the British programme, the United Kingdom was dependent on American superiority over the Soviet Union.

Even so, the United Kingdom continued to develop its own weapons, and its interest in heavy water from New Zealand soon took on a new, more sinister, dimension. On a visit to New Zealand in September 1952, John Cockcroft met with Cabinet and the Defence Science (Policy) Committee and made it clear that the British wanted heavy water not just to use as a moderator in atomic piles, but also from a 'defence research angle'.[18] While in New Zealand, Cockcroft also inquired as to the availability of bulk supply of electricity for use in a uranium enrichment plant.

In May 1953, the New Zealand Cabinet approved, in principle, the construction of a joint New Zealand–United Kingdom combined heavy-water and electricity generating plant. The focus now moved onto determining the economics of the project and the nature of the agreement between New Zealand and the United Kingdom. In late 1953, the British briefly withdrew from the project, when it seemed possible they could soon get heavy water from the United States at 'a keen price',[19] but in March 1954 the heavy-water project was revived. At a meeting of the British Chiefs of Staff, Sir Norman Brooks, Secretary to the Cabinet, reported plans to improve Britain's capacity to manufacture hydrogen bombs by obtaining thorium from South Africa and heavy water from New Zealand. The next week Marsden was advised that the United Kingdom might re-open discussions on the heavy-water project. Loath to put the reasons for the renewed interest in writing, Marsden cryptically and verbosely described it to DSIR secretary Hamilton, in New Zealand, as 'a very special urgent important reason'.[20] Whether or not Hamilton had any idea what Marsden was talking about is unclear. On 23 April 1954, Viscount Swinton, British Secretary of State for Commonwealth Relations, advised Holland that, on the basis of new cost and supply information, the United Kingdom Government now wanted to proceed with the heavy-water project, but this time attached great importance to maintaining secrecy. On the same day, Cabinet authorised Holland to tell the British High Commissioner that the New Zealand Government was willing to go forward with the proposed combined heavy-water and electricity plant in the Wairakei geothermal area.

While official reports did not state any link between weapons development and Britain's interest in New Zealand's heavy water, there was speculation that this was the case. Napier's *The Daily Telegraph* suggested in 1955 that 'the availability of heavy water (and hence of heavy hydrogen) from New Zealand was an important factor in the British Government's decision to make the H-bomb'.[21] Even so, according to Kevin Clements, 'no political party expressed moral or political scruples about the possible diversion of heavy water into weapons production'.[22]

When the British Cabinet formally decided to proceed with building a hydrogen bomb, they abandoned plans to use heavy water from New Zealand for the project, but, based on revised cost estimates for heavy water from America, they did want the heavy water for a nuclear reactor project. In announcing the re-establishment of the joint venture between New Zealand and the United Kingdom Atomic Energy Authority (UKAEA) — a new agency which had taken responsibility for British atomic energy projects away from the Ministry of Works — Holland told Parliament that New Zealand would be making a 'worthwhile and unique contribution not only to its own power resources, but also to the development of atomic energy, which, used for peaceful purposes, may well revolutionize the world's industrial processes'.[23] Geothermal Developments Ltd, whose shareholders were the New Zealand Government and the UKAEA, was formed in February 1955 to produce electricity and heavy water from Wairakei's geothermal steam. Marsden, who had retired from the public service in 1954 and returned to New Zealand, was appointed technical advisor to the board.

Design work proceeded to the stage where prices for equipment, materials and labour could be accurately estimated — but this doubled the cost of the plant, raising the cost of the heavy water it would produce from £44,000 to £90,000 per ton. In January 1956, the UKAEA advised that, faced with the projected price increases, they felt forced to withdraw from the project. Holland expressed regret at 'the abandonment of an interesting partnership agreement', but the project was still of value to New Zealand.[24] Holland decided that New Zealand would proceed with production of electricity from geothermal steam at Wairakei, and plans were revised to construct a larger power station to absorb the steam that would no longer be needed for heavy-water production. John Cockcroft sent a personal letter to Ernest Marsden, expressing his regret at having to abandon the heavy-water project. He added: 'I hope the New Zealand Prime Minister will not feel that he has been too badly let down in our first attempt at New Zealand–U.K. co-operation in the Atomic field. I hope we shall come back later with other projects which will be successful.'[25]

'A POLITICAL H-BOMB'

While New Zealand didn't have the 'great empty spaces' that Australia had, she did have many island territories, which Britain soon coveted for testing its hydrogen bombs. Scientists from the Aldermaston weapons development laboratory said the hydrogen bomb testing site should be at least 800 kilometres from inhabited land or shipping lanes. The best options were considered to be 'various remote islands or the icy wilderness of Antarctica'.[26]

British Prime Minister Anthony Eden was quick to ask for New Zealand's assistance with the testing programme, requesting help obtaining food, stores and fuel, use of the Auckland naval base for repairs, and the use of Penrhyn Island, in the Cook Islands, as a weather and radar station. New Zealand was advised the operation was top-secret. Holland gave the matter careful consideration and agreed to the British request, saying New Zealand would be 'very glad' to provide the facilities requested.[27]

William Penney, who ran the weapons development programme at Aldermaston, discussed the test site with the British Navy, which favoured the New Zealand sub-Antarctic island group, the Antipodes Islands — now a World Heritage site and home to protected species of albatross, penguins, petrels and seals — some 800 kilometres south-east of New Zealand. Here, a bomb could be detonated in a ship anchored close to the selected island. The Navy suggestion was rejected, however, and the United Kingdom chose the Kermadec Islands — another New Zealand territory, this time some 1000 kilometres north-east of New Zealand — as the most promising site. This group of volcanic islands is dominated by Raoul Island, and is now part of New Zealand's largest marine reserve. But in the 1950s, while the island was already designated a reserve for flora and fauna protection, there was little awareness of the Kermadec Islands' role as a breeding ground and safe haven for hundreds of marine and avian species: they were just another group of remote islands inhabited only by a small group of three meteorological staff and their support team of a carpenter, a mechanic, a cook and two farmers.

The decision whether or not to let the United Kingdom use the islands in this way was up to New Zealand Prime Minister Sidney Holland. In his *Dictionary of New Zealand Biography* profile of Holland, Barry Gustafson describes him as being a 'fervent admirer of Britain', and quotes Holland, whose parents were born in England, as describing himself as 'a Britisher through and through' who was determined to maintain links with Britain.[28] Some correspondence from 1955 shows, however, that regardless of the strength of Holland's determination to maintain links with Britain, his determination to remain in office, and even serve New Zealand's own interests, was stronger.

In May 1955, Eden made a direct request to Holland regarding the use of the Kermadec Islands for the British hydrogen bomb tests. Eden described how the weapons could be either exploded on one of the islands from a tower, or fired in a ship anchored near an island, and asked if Holland would agree in principle to the weapons trials so that the United Kingdom could investigate the site further. Eden concluded by expressing his earnest 'hope that, in the interests of our common defence effort and the importance of the deterrent for Commonwealth Strategy, you will find it possible to agree'.[29] Having not received a reply by the beginning of July, Eden sent a prompting message to

Holland, asking for a favourable reply to his May message, adding 'I am sure that we can count on your full co-operation in a project that is so important to the Commonwealth and the defence of the Free World'.[30]

Despite his 'fervent' admiration of Britain and desires to maintain links with the country, it seems that Holland was immediately wary of the British request, and took note of the negative publicity surrounding earlier newspaper reports of British plans to test nuclear weapons in Antarctica. He also sought the opinion of Marsden (still much sought-after, in retirement, for his expertise on nuclear and radiation issues), who advised Holland that, while an isolated island in the Pacific was 'a logical choice' for the proposed weapons test, the Kermadec Islands were not necessarily the best option. He acknowledged the weather was suitable, but noted the presence of occasional ships and aircraft in the area, and reminded Holland of the Japanese fishermen who had suffered radiation sickness after the United States' hydrogen bomb detonation at Bikini Atoll in March 1954. Marsden acknowledged that the Government might on the one hand feel a 'moral obligation' to co-operate with the British request, but on the other hand might 'feel that the sacrifice and difficulties in the use of the Kermadecs is questionable'.[31] Without bluntly advising Holland to refuse the request, Marsden suggested the Auckland Islands, some 320 kilometres south-south-west of New Zealand, as a preferable alternative to the Kermadecs.

Historian James Belich argues that New Zealand was an ideological and economic semi-colony of Britain until the 1960s. Holland, however, despite his deep affection for Britain, appears to have put his own and New Zealand's interests ahead of Britain's. His decision to refuse Eden's request was independent and pragmatic: the National Party had won the 1954 general election with only 43.8 per cent of the vote, and their continuance in power could not be taken for granted. On 15 July 1955, Holland warned the British High Commissioner in Wellington that the use of the Kermadecs for nuclear tests would be 'a political H-bomb' for New Zealand — not least because they would take place in an election year — and declined the British request.[32] Eden expressed his disappointment at Holland's refusal, reiterating the importance of the planned trials to the 'defence of the free world' and advising that if Britain did not find a suitable alternative he might be compelled to ask Holland to reconsider the matter.[33]

As part of the search for an alternative test site, Eden requested New Zealand's permission to investigate Penrhyn Island as an aircraft base and asked if the Royal New Zealand Navy survey ship HMNZS *Lachlan* would assist in a Pacific ground survey. Although Eden was still hoping that Holland might change his mind about the Kermadec Islands, a new site was chosen. Malden Island and Christmas Island, in the Northern Line Islands (now part of the Republic of Kiribati), were selected, and in early 1956 the HMNZS *Lachlan* surveyed the islands on behalf of the United Kingdom. Holland did not change his mind about the use of the Kermadecs, and in March Eden formally advised New Zealand that the range for the British hydrogen bomb testing programme would be Christmas Island and Malden Island.

While the public announcement that Christmas Island would be used for the bomb tests did not come until June, on 2 April 1956, following a press leak in the United Kingdom, New Zealand newspapers reported that the 'isolated coral atoll in the Pacific' would be the base for Britain's first hydrogen bomb explosion.[34] Christmas Island, the biggest coral atoll in the world, covered 642 square kilometres with its mixture of land and lagoons. There was already a runway on the island — it had been used as an American base during the Second World War — but the British plan was to refurbish the airbase and build the necessary accommodation and facilities. A new airstrip, a meteorological station and a tent camp would be set up at Malden Island.

There was, by now, opposition to the proposed hydrogen bomb tests. The New Zealand Labour Party, since its first term as government in 1935, had had a policy of independence in foreign policy, and this now applied to nuclear testing. In May 1956, when three Labour MPs asked Holland whether, in view of the conflicting reports on the likely effects of hydrogen bomb tests, he would protest at the continuation of nuclear bomb tests in the Pacific, Holland replied that 'the development of this branch of the nuclear sciences must continue' and 'periodic tests are essential to this work'.[35] In a later statement he added, 'New Zealand will be helping to ensure that the United Kingdom remains in the forefront in the field of nuclear research'.[36]

He saw continued testing as a positive move for enhancing Commonwealth, including New Zealand, security, but to maintain the support of the New Zealand public it became a case of 'not in my backyard'. On the one hand, he was faced with the public and political desire to maintain strong relations with the United Kingdom and with New Zealand pride in helping Britain's military and scientific achievements, and, on the other hand, concerns about radioactive fallout and damage to New Zealand territory. In this clear conflict between international loyalty and domestic politics, the latter won. But by providing logistical support for the British tests while withholding New Zealand territory, he tried to maintain a balance between the two.

Holland and his Government also showed assertiveness and independence with regard to a controversial United States proposal. As Malcolm Templeton reveals in *Standing Upright Here*, in late 1956 New Zealand firmly rejected a United States claim to sovereignty over Penrhyn Island, which had become strategically important to the Pacific nuclear-testing programme, pointing out that the island had been New Zealand territory for more than 50 years. The American claim was eventually dropped.

In response to further British requests, New Zealand provided facilities for a weather station and a radio station on Penrhyn Island, and the RNZAF transported several officials to the test zone. Then, in late March 1957, HMNZSS *Pukaki* and *Rotoiti* joined the British squadron at Christmas Island. The frigates were responsible for 'air/sea rescue, anti-submarine watch, thermal flash monitoring and water sampling to test for radiation contamination', but their main role was to collect meteorological data in and around the test zone.[37] As Maurice Hayman, a radio operator on HMNZS *Pukaki*, later recalled:

> A canvas hut was erected on the stern, where the balloons were inflated, and then brought out for release, once the transmitter had been attached. The balloon was released at a size of about six feet across. After rising to heights of 50,000 feet or more, it would by then be about the size of a two storey house, before finally bursting. The idea was to monitor the upper levels, forecasting wind speed, direction, temperature, humidity and precipitation up to 100,000 feet as part of an overall weather forecasting system, prior to each burst.[38]

LET THEM 'EAT ONLY COCONUTS'

While the New Zealand Government was happy to offer logistical support to the British tests, there was growing public concern — in New Zealand and in the South Pacific in general — about the effect of radioactive fallout from the

Opposite: The main role of HMNZSS *Pukaki* and *Rotoiti* was to gather meteorological data in and around the nuclear test zone. Weather balloons were inflated in a canvas hut on the ship, then released: as they rose, the balloons would gather data on wind speed, direction, humidity and precipitation. Photographer Neil Anderson, Royal New Zealand Navy Museum, AAO0037.

tests. In response to concerns passed on by the Minister of Island Territories and a Samoan petition to the United Nations, Deputy Prime Minister Keith Holyoake advised the New Zealand public that the British Government had assured New Zealand there would be no radioactive hazards to the inhabitants of the Cook Islands, the Tokelau Islands or Western Samoa. If Britain had made the assurance, it seemed, it was good enough for New Zealand. In April 1957, the monthly review journal *Here & Now* reported 'there was, in the way Mr Holland solemnly relayed a British assurance that no inhabited islands would be affected by radiation, almost a tone of pride in assisting at headline-making events'.[39]

The same month, Harold Turbott, Deputy Director-General of Health and chairman of the Radiological Advisory Council charged with advising the minister on radiation issues, responded to an External Affairs request for information about the risk to people in New Zealand's Pacific Islands territories. Turbott said that New Zealand 'must accept the British Government's assurance that there is no danger to our Island Territories', but he did suggest that radiation monitoring in the region be extended.[40] The British assurances might have been passed on to the public, but officials did not necessarily accept them. In response to the External Affairs query, the Department of Health acknowledged that fish could be vulnerable to radioactive contamination and suggested 'the inhabitants might be advised to eat only coconuts during the period of the tests'.[41]

Even though New Zealand gave considerable logistical support to the United Kingdom, Holland was not given advance notice of the first British hydrogen bomb test, which took place in the atmosphere above Malden Island. It came as a 'complete bombshell' to Holland who said his first news of the bomb came from the newspaper.[42] Templeton describes Holland as reacting to the news with 'great surprise and anger';[43] he wrote to British Prime Minister Harold Macmillan to complain about the lack of notice and advise of the concerns of the New Zealand public about the risk of radiation hazards, to which Macmillan replied with what Templeton describes as acute annoyance. There had been hints that a test would be taking place soon, and three New Zealanders were official observers of the tests.

Opposite: The men on HMNZS *Pukaki* 'had a good view' of the first nuclear explosion of Britain's Operation Grapple and were 'much impressed' said Prime Minister Sidney Holland. The men were ordered to put on protective clothing and stand or sit on deck for each bomb detonation. Photographer Neil Anderson, Royal New Zealand Navy Museum, AAO0012.

On the Royal Navy ship HMS *Alert*, along with two military observers was Bert Yeabsley, deputy director of the Dominion X-Ray and Radium Laboratory. Wearing their white anti-flash suits, hoods, glasses and dark goggles, the observers sat on the ship's deck facing away from the blast, turning towards it after ten seconds. Yeabsley reported being impressed by the way the British were conducting the tests, concluding that the tests 'were being made in such a fashion that the possibility of highly active local fall-out was reduced to a minimum and that no person under the care of the New Zealand Government was liable to suffer radiation damage from the operation'.[44]

Britain's Operation Grapple, the codename for the project to test hydrogen bombs in the Pacific, began with three prototype weapons tested in May and June of 1957, detonated at a height of about 2400 metres, in the atmosphere above Malden Island. Further test series — codenamed Grapple X, Grapple Y and Grapple Z — took place off the coast of Christmas Island in November 1957 and from April to September 1958, with yields peaking at 3 megatons for an explosion on 28 April 1958.

OPERATION GRAPPLE ATMOSPHERIC TESTS, 1957–58

LOCATION AND DATE	ESTIMATED YIELD	RNZN POSITIONS
Grapple series, above ocean near Malden Island Short Granite, 15 May 1957 Orange Herald, 31 May 1957 Purple Granite, 19 June 1957	300 kt 700–800 kt 200 kt	93–278 km from ground zero
Grapple X, Y, Z series, above ocean near Christmas Island Blue Danube, 8 November 1957 Grapple Y, 28 April 1958 Pennant, 22 August 1958 (balloon burst over land) Flagpole, 2 September 1958 Halliard, 11 September 1958 Burgee, 23 September 1958	1.8 Mt 3 Mt 24 kt 1.2 Mt 800 kt 25 kt	37–278 km from ground zero

Note: kt = kiloton and Mt = megaton: units of explosive force equivalent to 1000 and 1 million tons of TNT, respectively.

Source: Compiled by the author from: Lorna Arnold, *Britain and the H-bomb*, Palgrave Macmillan, Houndmills, 2001; John Crawford, *The Involvement of the Royal New Zealand Navy in the British Nuclear Testing Programmes of 1957 and 1958*, New Zealand Defence Force, Wellington, 1989.

Following the first explosion, on 15 May 1957, *The Dominion* editorial noted that Britain had established 'her credentials as a member of the thermo-nuclear club'.[45] In a radio broadcast a week after the test, Holland assured the public that 'the tests are being carried out with the utmost care and regard for the safety of human life', going so far as to say 'British scientists seem to have mastered the problem of dangerous fallout associated with previous nuclear tests', although this was probably more a factor of it being a relatively small blast than the British having any control or 'mastery' over the fallout. Holland also said that the *Pukaki*, stationed some 50 nautical miles from the explosion, reported 'that she had a good view and was much impressed'.[46]

The truth was that — despite Yeabsley's and Holland's apparent confidence in the way the tests were being run — these weapons were so powerful and so unpredictable that there was no way any of the politicians or scientists could have 'known' they would be safe. What's more, they were generally assessing safety on the one-off risk of each test. The cumulative effects of radioactive fallout produced by atmospheric nuclear bomb tests would soon be showing up in the bones and teeth of children throughout the world.

Roy Sefton, a radio operator stationed on the *Pukaki*, recalled the first blast:

My eyes were closed behind dark glasses, but opened at the flash — and I saw my finger bones. Then there was a rumbling like stampeding horses before the shock wave hit the ship.[47]

Just 15 seconds after the explosion, *Pukaki*'s crew were ordered to 'open your eyes, stand up and face the burst'. Commander Hale, in his report on the explosion, described the view:

. . . The fire ball grew in size shaped like a round firy [*sic*] and turbulent cauliflower changing from an angry deep red streaked with grey to a larger smouldering ball of cloud with a glowing centre . . . . Between the 2nd and 3rd minutes the terrific up draught of air and cloud soon became apparent by what appeared to be a strikingly white water spout being drawn into the centre of the fire ball, this rising mass increased in volume until the more familiar but equally fantastic shape of the mushroom was evident to everyone.[48]

Maurice Hayman, a telegraphist on the *Pukaki*, recalled a later blast:

> The weather was calm, and I sat down on the quarterdeck dressed in longs,
> shirt and anti flash gear. The officer in charge gave the order to face to
> starboard, and we could hear the line countdown on the ship's speakers.
> With 15 seconds to go we were given the order to close and cover our eyes.
> As it detonated, I felt the heat and saw the flash through my eyelids and
> fingers. About half a minute later we were told to face the burst. All cloud
> cover had gone, and the coloured fireball was climbing rapidly upwards.
> It was an impressive sight . . .[49]

Following the first test, the *Pukaki* sailed within 6 nautical miles of ground
zero to rendezvous with a British ship, although Geiger-counter measurements
recorded no noticeable radiation in the air or water.

All the ships involved in Operation Grapple, however, took protective
measures against the possibility of radiation contamination. Parts of the
ships could, if necessary, be sealed off to provide an airtight enclosure, and
the ships contained wash-down equipment to prevent any unexpected fallout
from settling and to wash away any that did. Radiation detection systems
were in place on all ships, and the British authorities on Christmas Island
supplied protective clothing.

Maurice Hayman, recalling one of the later blasts in the series, said:

> It wasn't clear at that time why everyone not required for duty below had
> to sit up on deck close to the explosion. The correct procedure for the fleet
> in the event of a Nuclear attack, was to spread out far and wide, close the
> ship down, and start up outside water sprinklers. A released British Defence
> Research Policy Committee report dated 20th May 1953, on the Nuclear
> Weapon Trials, answered this question. It read 'The object of the exercise, is
> to discover the detailed effects of various types of explosion on equipment,
> stores, and men with and without various types of protection.' That explains
> why after the last balloon release, we were not shut down and heading flat out
> away from the burst.[50]

Opposite: British authorities supplied protective clothing to all the men on the *Pukaki*
and *Rotoiti*, and radiation testing took place after each of Britain's nuclear explosions.
Photographer Neil Anderson, Royal New Zealand Navy Museum, AAO0029.

Prime Minister Sidney Holland (right) and his deputy, Keith Holyoake, in August 1957, a month before Holland retired and Holyoake became prime minister. Even as Holland's deputy, Holyoake questioned the value of nuclear weapons testing. Photographer Morrie Hill, 1/2-177291-F, Alexander Turnbull Library, Wellington, New Zealand.

Holland retired as prime minister in September 1957, and was succeeded by Keith Holyoake. As deputy prime minister, Holyoake had already been vocal on nuclear issues, issuing a statement earlier in September, following a visit to New Zealand by the British Minister of Defence, 'that New Zealand would not become an atomic power: it would acquire neither atomic nor nuclear weapons and would not become a base for their storage'.[51] Holyoake was more personally against nuclear testing than Holland was, and in 1958, when he agreed to a British request for New Zealand assistance with the second round of British tests, he ended his letter to Macmillan by commenting that a question increasingly being asked by the average citizen in his part of the world was: 'Why if there is no danger from these tests, do the British and Americans not hold them nearer home?'[52]

Labour, who won the 1958 general election, had campaigned on opposing all future nuclear testing. The new prime minister, Walter Nash, was British — he had been born in England and came to New Zealand as an adult — and Templeton has described him as always reluctant 'to come out in direct opposition to what was presented as a vital British interest'.[53] Despite the election promises, Nash agreed to fulfil the National Government's earlier undertaking to support the 1958 British test series, and the *Pukaki* and *Rotoiti* offered similar support to that given in 1957. Labour's stated opposition to nuclear testing did not extend to opposition to other nuclear technologies. In 1959, following the 1955 discovery of uranium deposits on the West Coast, the Minister of Mines signed a secret agreement in which the UKAEA funded a drilling programme to investigate the region's uranium deposits. And it was the Labour Party that advocated a North Island nuclear power station when the National Government was pushing for a Cook Strait cable to carry electricity from the South Island to the North Island. If it was in New Zealand's financial interests, and would help the country become more economically independent, then Labour supported it.

New Zealand had minimal links to the United States testing programme. In 1957, the New Zealand deputy defence attaché, based in Washington, attended an American nuclear test at Nevada as part of a group of 80 invited observers from countries with defence links to the United States. Again, a New Zealander was invited to and observed an American test at Enewetok Atoll in July 1958. The New Zealand Chiefs of Staff had noted that the opportunity for New Zealand officers to witness nuclear explosions and study their effects first hand occurred very infrequently, and the opportunity would be of considerable benefit to the officer involved and the people he could brief about it. The observer was not of the same opinion, however. After the test, he reported that the limited information provided by the American authorities meant that observation of the test was of little value to a New Zealand observer. No New Zealand military observed any more American tests.

OPPOSITION TO NUCLEAR TESTING

The possibility of nuclear war was now becoming part of popular culture, and books like Nevil Shute's 1957 novel *On the Beach*, and the subsequent film of it in 1959, had a powerful effect on the public imagination. The story was set in 1963 Melbourne, as people in Australia, New Zealand and South Africa waited for global circulation currents to deliver a deadly cloud of radioactive fallout from a northern hemisphere nuclear war. The fact that the story was set in the southern hemisphere, and featured the impact of a northern hemisphere war on countries like New Zealand — death would come, but it would be slower and with more warning — was a chilling prospect to many New Zealand readers and helped strengthen the anti-nuclear movement.

In New Zealand, organised anti-nuclear sentiment had begun almost immediately after the end of the Second World War. New Zealand's first Hiroshima Day march was in Christchurch in 1947. Nuclear weapons testing by the United States, the United Kingdom and the Soviet Union strengthened anti-nuclear resolve, and an international anti-nuclear movement emerged in opposition to all nuclear weapons. In 1950, the World Peace Congress started collecting signatures worldwide for the Stockholm Peace Appeal, which stated:

We demand the absolute banning of the atomic weapon, arm of terror and mass extermination of populations [*sic*]. We demand the establishment of strict international control to ensure the implementation of this banning measure. We consider that any government which would be the first to use the atomic weapon against any country whatsoever would be committing a crime against humanity and should be dealt with as a war criminal.[54]

The appeal eventually collected 650 million signatures worldwide, more than 20,000 of them from New Zealand. Despite this level of support, anti-bomb protesters were often accused of having Communist sympathies, and in 1950 Peter Fraser, prime minister of the 1940–49 Labour Government, described the Stockholm Peace Appeal as 'just another Soviet weapon'.[55]

The anti-nuclear movement was fuelled throughout the 1950s by nuclear accidents, like those at Chalk River in Canada in 1952, and at Windscale in the United Kingdom in 1957, and by growing fears about genetic damage caused by radioactive fallout. The Chalk River meltdown was the world's first serious nuclear reactor accident. A power surge and subsequent loss of coolant led to the meltdown of the NRX reactor (the research reactor that four New Zealand scientists had worked on), with hydrogen explosions damaging the surrounding buildings, and radioactive material being released into the atmosphere and the cooling water. At Windscale — a plant producing plutonium and tritium for British nuclear weapons — a four-day reactor fire led to the release of radioactive material across the United Kingdom and Europe, and the permanent shutdown of the reactor.

In a 1958 *New Zealand Listener* editorial, Monte Holcroft said that, in spite of all the expert opinions and assurances, a suspicion was growing

In 1950, former Labour prime minister Peter Fraser described the Stockholm Peace Appeal — which aimed to ban atomic weapons, and which 20,000 New Zealanders signed — as 'just another Soviet weapon'. PAColl-8211, Alexander Turnbull Library, Wellington, New Zealand.

that 'the scientists are using forces they do not fully understand'.[56] While these events raised fears about the safety of nuclear power, the British and American bomb tests in the Pacific brought the nuclear world closer to New Zealand and gave impetus to the country's nascent anti-nuclear movement. There was widespread opposition to the tests in the Pacific — from church groups, students, women's organisations, Maori, unions and scientists — with many groups passing anti-nuclear resolutions and many groups and individuals writing to Holland to voice their opposition, particularly after the first hydrogen bombs were tested. The strongest objection to bomb tests arose from fears about the genetic and health effects of radioactive fallout.

To people calling for an end to the testing of hydrogen bombs, Holland parroted Churchill's reply to his own voiced concerns: that it would be 'a great disservice to the free world' for Britain to seek to 'impede the progress of our American allies in building up their overwhelming strength in a weapon which provides the greatest possible deterrent against the outbreak of a third world war'.[57] Australia and New Zealand followed Britain's lead on this issue, while newly independent India called for an immediate halt to all nuclear testing.

Holland responded to New Zealanders' opposition to bomb testing by trying to justify Britain's role in the arms race as essential, and by minimising the harmful effects of fallout. In a radio broadcast that followed the explosion of the first British hydrogen bomb, Holland talked about the many letters he had received from around New Zealand, both for and against the nuclear tests, in which some people called for the Government to oppose all nuclear testing. While he believed that most of these views were 'sincerely held', Holland also cautioned that some of the letters were part of a programme of 'Communist propaganda' against nuclear testing. Holland defended his Government's support for Britain's nuclear testing by referring to the 'great deal of study of scientific information' they had conducted, and to the vulnerable position of the 'Motherland', which must show 'those who need to be shown that she has the means to defend herself'.[58]

But while there were mixed views about the ethics of bomb testing, concern about the health effects of fallout was ubiquitous. Holland described people's major concerns being 'that the atmosphere is being poisoned, that food supplies will be contaminated, and that flesh and blood itself are being attacked by unseen, deadly radiations'. In an attempt to reassure listeners, Holland pointed out that the amount of radiation people were exposed to as a result of nuclear testing was small compared with radiation from cosmic rays or x-rays. His assurances, however, went beyond what he could know from current scientific information. To people concerned about the impact of fallout on Pacific food sources, he said the British tests were 'high air bursts and the fallout will be negligible and will filter back from the stratosphere without doing harm'. While voicing his support for the eventual banning of nuclear testing, Holland stated that 'the course being followed by Britain is the right course, and we must continue to support her. Her aim is the security of the Commonwealth and the free world and our safety lies in that security.'[59]

Throughout the 1950s, New Zealanders continued to present petitions asking the New Zealand Government to do more to protest against atmospheric and underground nuclear tests. The Government's response depended on which party was in power. While Holland's National Government voted against a 1956 United Nations resolution calling for a

World Court opinion on the legality of atmospheric testing, in 1959 Nash's Labour Government supported a United Nations resolution condemning nuclear tests, sought a nuclear test ban treaty, and helped develop the world's first nuclear-weapons-free zone, in Antarctica. After regaining power in 1960, however, Holyoake's National Government voted against a United Nations resolution that declared the use of nuclear weapons as contrary to the laws of humanity.

In spite of growing opposition to nuclear testing, the arms race continued. In 1957, the same year that Britain was testing its first hydrogen bomb, the Soviet Union was launching the world's first intercontinental ballistic missiles, and, later that year, Sputnik 1, the first artificial satellite. These developments raised the spectre of intercontinental nuclear missiles, or nuclear weapons launched from space. The Soviet president, Nikita Khrushchev, wanted the West to be fearful of the Communist superpower. As John Lewis Gaddis has written, from 1957 through 1961 Khrushchev 'openly, repeatedly, and bloodcurdlingly threatened the West with nuclear annihilation'.[60] In turn, Khrushchev's threats were used as justification for the United States continuing its nuclear-testing programme.

'A VISION OF A MAN-MADE APOCALYPSE'

In response to anxiety about fallout, the United States and the United Kingdom agreed in 1958 to suspend nuclear testing for a year. Britain never resumed nuclear testing, but in 1962 the United States began a new Pacific test series that led to increased anti-nuclear sentiment in New Zealand. By this time France had become the world's fourth nuclear power, after testing nuclear bombs in French territory in North Africa. Before the 1962 tests, Holyoake alerted the United States Government to the likely outcry in New Zealand and the Pacific. To the New Zealand public he issued a press statement blaming the resumption of testing on the Soviet Union: its testing a 50-megaton bomb in the atmosphere in 1961 had forced the decision upon the West, he said.

The United States' 9 July 1962 bomb test was a pivotal experience for many New Zealanders. The 1.2-megaton hydrogen bomb was launched on a rocket and exploded 320 kilometres above Johnston Atoll. The high-altitude test, designed to test the effect of a nuclear explosion on radio- and radar-communication, disturbed New Zealand's telecommunications systems and created an artificial aurora, described in *The New Zealand Herald* as an intense glow above the northern horizon that spread rapidly across the sky before 'the luminous red band widened, and quivering white shafts of light could be seen within it'.[61] *The New Zealand Herald* editorial the next day described the eerie glow from the nuclear explosion as doing 'more than a hundred protest marches to fill men's minds with dread'.[62] David Lange, who would be prime minister of a Labour Government from 1984 to 1989, later recalled that 'the confusion in the sky that night haunted me as a vision of a man-made apocalypse, a terrifying retaliation of natural forces against the evil of unnatural invasion and a warning that a small country at the edge of the world in the South Pacific was no longer far enough away from the quarrels of the great powers to escape their consequences. It was a shock to realise that the power of nuclear weapons could straddle the world and unleash a threat on an inoffensive country like New Zealand.'[63]

The American tests created mixed feelings in New Zealand. The test was shocking, but the United States was still an ally and the leader in the West's fight against Communism. While *The New Zealand Herald* acknowledged that the 'yearning to see these dreadful engines of destruction abolished must be nearly universal', it also stated that to 'clamour for immediate nuclear disarmament flies in the face of reality'.[64] Later that year, the New Zealand Atomic Energy Committee, an advisory committee established in 1959, responded to a request from the Prime Minister's Department by commenting on the potential impact of high-altitude nuclear tests on the ionosphere. They advised that such tests would increase the levels of long-lived radioisotopes such as strontium-90, and recommended that the New Zealand Government 'oppose further high-altitude tests unless such tests have the support of and the scientific observations are coordinated by an internationally recognised scientific organisation'.[65]

THE SCIENTISTS RESPOND

The post-war period saw proposals for a research reactor in New Zealand, a search for uranium on the West Coast, nuclear testing in the Pacific, and the first reports of fallout deposition in New Zealand and the Pacific. Politicians were driving decisions about New Zealand's level of support and involvement

Sir Ernest Marsden in June 1961, on board the *Sydney Star* at Bluff, New Zealand, testing the radioactivity of a sample of seawater. 'I wish I could start my career again and work on these radiobiological problems,' he told a colleague. F-153607-1/2, Alexander Turnbull Library, Wellington, New Zealand.

in these ventures, but what did New Zealand scientists — particularly those working in the field of nuclear physics — make of these developments? It is interesting to look at the example of Ernest Marsden, New Zealand's original nuclear advocate, because of the way that his attitude to nuclear weapons and nuclear technology changed over the decades after the Second World War.

At about the same time that he was advising Holland against allowing the United Kingdom to test hydrogen bombs in the Kermadec Islands, Marsden was beginning his own research into the biological effects of background radiation. In his 'retirement', which began in 1954, Marsden worked up to six days a week, either from the attic laboratory at his home, or as a guest worker at the DSIR's Dominion Physical Laboratory or the Royal Cancer Hospital in London. He was passionate about this new line of work, telling a colleague: 'I wish I could start my career again and work on these radiobiological problems.'[66]

Marsden liked an audience and received a lot of press coverage. He sometimes talked up the effects of radiation from bomb tests and sometimes minimised them, pointing out that radiation levels from fallout were very low in comparison to natural background radiation. He consistently, however, said that the effects of radiation from bomb fallout were not fully understood and deserved further study.

Marsden began speaking out against nuclear weapons testing in 1959, after the United Kingdom had completed its nuclear-testing programme in Australia and the Pacific. He highlighted the worldwide increase in radioactive fallout resulting from Soviet and American nuclear tests, and told the *Auckland Star* 'the time has come for an absolute standstill on such atomic explosions to give time for a proper assessment of the damage already done to us and to our children even yet unborn'.[67]

As we have seen, this wasn't the first time that Marsden had publicly opposed nuclear weapons. After the end of the Second World War, he was one of many international scientists to support the Baruch Plan, which had stipulated that the United States dispose of its atomic weapons, stop all weapons work and turn over its atomic energy knowledge to the United Nations. In a 1947 speech, Marsden had said that it was not safe to develop atomic energy until there was a practical and enforceable agreement that it would not be used for atomic bombs. At the same time, however, he promoted the development of a nuclear reactor in New Zealand as being of defence significance to the Commonwealth. And later, in the early 1950s, he supported British plans to develop nuclear weapons, and was keen for New Zealand to assist the British nuclear programme by constructing a nuclear reactor and providing heavy water and uranium.

After the British nuclear programme ended in 1958, Marsden declared that New Zealand was partly to blame for the Commonwealth 'falling miserably behind in nuclear development'. If there was a third nuclear power, he said, there would be no 'bombing competition' between Russia and America.[68] Marsden continued to criticise New Zealand's lack of investment in defence science, including telling Holyoake that New Zealand had been 'grossly discourteous and negligent of opportunities to help Britain' in this area; a reference to New Zealand's continued failure to construct an atomic pile.[69]

Why was Marsden speaking out against nuclear weapons at the same time as implicating New Zealand in the United Kingdom's failure to keep up with the arms race? As journalist Tony Reid described in a newspaper profile of Marsden, his attitudes to nuclear weapons development were 'ambiguous and sometimes contradictory'.[70] It is possible that, despite his initial personal misgivings about the post-war development of nuclear weapons, Marsden's loyalty to Britain, along with the close involvement of many of his friends and former colleagues in the British nuclear programmes, caused him to push these misgivings aside. Marsden was easily seduced by science — as demonstrated by his offer in early 1945 to leave his position as head of the DSIR to take a scientific position on the North American nuclear programme — and the development of nuclear weapons was at the forefront of scientific and technological development. Once the British nuclear-testing programme was concluded, however, and with evidence of increased environmental radioactivity resulting from bomb fallout, Marsden had no hesitation in publicly opposing nuclear weapons.

Other scientists also had changing attitudes to nuclear issues. Jim McCahon — who had worked on the South Island uranium survey in the late 1940s, then had been seconded to the Atomic Energy Research Establishment at Harwell for two years before joining the Dominion X-Ray and Radiation Laboratory in 1951 — continued working in radiation physics but became a lifelong supporter of the Campaign for Nuclear Disarmament. Some scientists involved in the Manhattan project, like Maurice Wilkins, were so appalled by the bomb that they turned away from physics. (In Wilkins's case it was a great move for science, as he went on to share a Nobel Prize in Physiology or Medicine for his work on the structure of the DNA helix.)

After making a number of anti-nuclear statements to the media from 1959 onwards, Marsden began communicating his anti-nuclear weapons sentiments to Holyoake in 1961. When the French announced their proposal to move their nuclear test site to the South Pacific, Marsden advocated, in a 1963 letter to Holyoake, a nuclear-bomb-free southern hemisphere. He pointed out that fallout from nuclear bomb tests had so far had more of an impact on the northern hemisphere than the southern, and called on Holyoake to announce that New Zealand would not provide any assistance to countries carrying out bomb tests in the southern hemisphere, and suggested he call on other southern hemisphere countries to do the same. In May 1963 the New Zealand Government formally protested to the French Government over their preparations for a nuclear test at Gambier Islands. Later that year New Zealand was the fourth country — after the United States, the United Kingdom and the Soviet Union — to ratify the Partial Test Ban Treaty, demonstrating, in Holyoake's words, New Zealand's desire 'to see an end to nuclear tests that are likely to give rise to contamination of the atmosphere'.[71] The signatories to the Partial Test Ban Treaty agreed to prohibit, prevent and not carry out nuclear tests in the atmosphere, in outer space, or under water. Although it dealt to fears about fallout, it did not mean an end to the further development of nuclear weapons — nuclear-testing programmes just moved underground.

While focusing on his research into environmental radioactivity, Marsden continued to speak out against nuclear weapons development and testing. On a visit to South Africa he described the hydrogen bomb as 'the most striking example of the possibilities of misuse of modern scientific knowledge'.[72] In June 1965 he told *Salient*, the Victoria University student newspaper, 'we must do what we can to stop nuclear warfare. We must do what we can to promote nuclear disarmament'.[73] New Zealand's nuclear advocate had become a voice of caution.

The 1950s might have been economically prosperous for New Zealand, but nuclear bombs were being tested nearby and the country's milk and meat were becoming tainted with radioactive strontium: New Zealand's welcome to the 'atomic age' was radioactive fallout. Still, anti-nuclear sentiments were restricted to opposition to nuclear weapons, and did not extend to opposition to peaceful applications of nuclear technology, such as the emerging plans for a uranium mining industry to earn export dollars and fuel the nuclear power stations planned for New Zealand's future.

4

*

Uranium fever!
Uranium prospecting on the West Coast

*I believe that it is possible we have discovered the second most
highly concentrated uranium deposit in the world.*
— FREDERICK CASSIN, NOVEMBER 1955, BULLER GORGE[1]

It looks as if they've got something all right. — LES GRANGE,
DIRECTOR OF NEW ZEALAND GEOLOGICAL SURVEY[2]

As you drive up Buller Gorge, about a kilometre downstream from
Hawks Crag, a signpost marks Uranium Point, the site of New
Zealand's first uranium discovery. The bush-clad peaks on either
side of the road — Mt Cassin and Mt Jacobsen — are named after the two
sprightly septuagenarians who chanced upon the radioactive element one
Sunday afternoon in November 1955.

The two veteran prospectors, Frederick Cassin and Charles Jacobsen,
were returning home after a day in the hills above the Buller Gorge. They
were searching for mica, but, with a copy of the DSIR Geological Survey
booklet *Prospecting for Radioactive Minerals in New Zealand* in hand, were
also on the lookout for uranium. As the story goes, they finished their day
with a few drinks in the Berlins Hotel then, needing to relieve themselves
on the drive to Westport, pulled their truck over to the side of the road.
Jacobsen placed their Geiger counter on the rock face, where for the first time

Frederick Cassin and Charles Jacobsen holding a Geiger counter to a sample of uranium-bearing rock.
The Dominion Collection, F-144916-1/2, Alexander Turnbull Library, Wellington, New Zealand.

that day the counter began an excited clicking; the highly radioactive rock sent the needle off the scale. The jubilant pair returned to the Berlins Hotel for the night, stopping the next morning to gather rock samples to take to Wellington.

Cassin and Jacobsen arrived in the capital on 9 November, armed with their samples of radioactive rock. Their find excited the national and international media, along with other prospectors, especially when Cassin announced that the reef where they found the rock had a radioactive count higher than at Rum Jungle in Australia's Northern Territory or the world's richest uranium mines in the Belgian Congo. 'I believe that it is possible we have discovered the second most highly concentrated uranium deposit in the world,' he said.[3]

In Wellington, Cassin and Jacobsen met Prime Minister Sidney Holland, who tested their sample with a Geiger counter and congratulated them on their initiative. While the rock samples were being analysed, 70-year-old Jacobsen entertained friends and fellow prospectors by holding the Geiger counter against a sample of the yellowish, black-streaked rock, and handing them the earphone. The agitated clicking of the counter and the sight of

Charles Jacobsen entertains the ladies with a sample of radioactive rock and a Geiger counter outside the Berlins Hotel in Buller Gorge, near where he and Frederick Cassin found uranium-bearing rock. *Weekly News*, 23 Nov. 1955. Courtesy APN New Zealand.

the indicator needle creeping around to 450 counts per second unfailingly impressed people.

A few days later, the DSIR's Geological Survey announced that the samples tested contained 0.27 per cent uranium: they were more than twice as uranium-rich as Australia's Rum Jungle ore. This was hugely significant. Discovery of the Rum Jungle deposit had led to a ten-year contract to supply uranium oxide concentrate to the British–American Combined Development Agency, a joint United States and United Kingdom agency formed to purchase uranium. If the prospectors had discovered an economically viable uranium deposit in New Zealand, the whole country would benefit.

Cassin and Jacobsen returned to the Buller Gorge to stake out and register their claim. Following the vigorous outdoor work they returned to the Berlins Hotel, where they were so busy enjoying their drinks and recounting tales of their find that a Westport man, James Fair, registered New Zealand's first uranium claim. On Monday, 14 November 1955, Fair registered a claim for a

400-hectare block on the south bank of the Buller River on each side of Batty Creek, surrounding but excluding any mineral lease applied for by Cassin and Jacobsen. Cassin and Jacobsen were only minutes behind Fair, however, and filed their claim the same day.

Les Grange, head of the Geological Survey, visited the Buller Gorge site a week after Cassin and Jacobsen's find, firing up the nearby town of Westport. *The Press* reported his visit:

> Quickly assembling a high-fidelity field-rate meter made at the Harwell Atomic Research Station in England, Dr Grange went immediately to the face of rock which the discoverers exposed when they made the find. 'That looks pretty good', he said as the needle showed a strong reading on the scale of his instrument. 'Hector, it's gone right off the scale!' exclaimed Dr Grange a few moments afterwards, when testing the same roadside rock face a few feet to the right.[4]

Grange estimated the reef to be 900 feet long, 300 feet deep, with an average width of about 6 feet (approximately 275 metres by 90 metres by 2 metres). Although these dimensions were yet to be verified, Grange said the deposit was a highly worthwhile mining proposition. 'It looks as if they have got something all right,' he said after visiting the site.[5]

Media reports of Cassin and Jacobsen's find gave other prospectors clues on where to look, and in the days that followed the West Coast swelled in the grip of a uranium rush. Ninety years after the gold rush that had led to the European settlement of this rugged area, the hills and river valleys were again alive with prospectors, this time eager to find uranium to fuel the United Kingdom and United States nuclear programmes. A uranium strike was also an exciting possibility domestically: a local uranium source would, it was thought, make nuclear power all the more affordable for New Zealand.

EARLY SURVEYS OF RADIOACTIVE MINERALS

New Zealand's first attempt to find uranium was Ernest Marsden's wartime survey of the fiords and beaches of the South Island, which found uranium in beach sands but did not reveal any promising sources of uranium for mining. Through the 1950s, global demand grew for uranium to fuel the developing nuclear weapons and energy programmes in the United States and

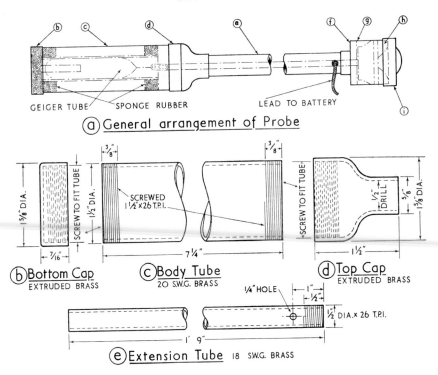

(a) General arrangement of Probe

GEIGER TUBE — SPONGE RUBBER — LEAD TO BATTERY

(b) Bottom Cap
EXTRUDED BRASS

(c) Body Tube
20 S.W.G. BRASS

(d) Top Cap
EXTRUDED BRASS

(e) Extension Tube 18 S.W.G. BRASS

Uranium prospecting was promoted as a weekend or summer hobby. The head of the Geological Survey, Les Grange, advised that: 'A Geiger counter can be made at home at an expenditure of little more than £10, and thus it is well within the reach of anyone who has ability in mechanical and electrical work. If a significant deposit of radioactive minerals is not found the small financial outlay will not be grudged, and, after all, it will have added interest to the summer holiday.' Source: Les Grange, *Prospecting for Radioactive Minerals in New Zealand*, DSIR Information Series No. 8, Government Printer, Wellington, 1954.

the United Kingdom. World production of natural uranium was estimated at 10,000 tonnes per annum, with the United Kingdom Atomic Energy Authority (UKAEA) importing uranium mainly from the Belgian Congo, South Africa, Australia and Portugal, and the United States Atomic Energy Commission (USAEC) sourcing its uranium domestically and from Canada. By 1955, Australia had uranium mines at Radium Hill in South Australia, Rum Jungle in the Northern Territory, and Mary Kathleen in Queensland, with contracts to supply uranium oxide to the UKAEA and the Combined Development Agency.

The UKAEA and the Atomic Energy Commission of Australia sent their chief geologists to New Zealand in 1952, who reported that it was unlikely there were economic uranium deposits in New Zealand. American geologists were more optimistic. The Vitro Organization (a large engineering company involved in the construction and operations programmes of the USAEC) advised New Zealand that the USAEC would look with 'considerable favor' upon the efforts of any friendly countries to uncover new uranium deposits, and said New Zealand would be able to negotiate 'very attractive contracts' with the United States.[6] On a 1954 visit to Australia, Grange met two geologists from the USAEC, Philip Dodd and Frank Frankovich, and invited them to New Zealand to help plan a search for uranium. The American geologists were working with the Australian Federal Government, helping to develop the Rum Jungle uranium deposit. They accepted Grange's invitation and visited New Zealand to help the Geological Survey plan a new approach to a uranium search.

Following the USAEC geologists' advice, in December 1954 the Geological Survey enlisted the help of amateur and weekend prospectors in a nationwide search for uranium with the publication of *Prospecting for Radioactive Minerals in New Zealand*. The Geological Survey booklet provided information on the field properties of radioactive deposits, along with instructions on likely places to find them and, for those unable to buy a Geiger counter — available for £40 each from local electronics suppliers — simple instructions on how to make one.

While Grange advised that the likelihood of finding economic uranium deposits in New Zealand was slim, the potential prize was great. The United Kingdom and the United States were eager for new uranium deposits to develop, and the New Zealand Atomic Energy Act of 1945 promised the first finder of uranium a financial reward.

URANIUM FEVER ON THE WEST COAST

When Cassin and Jacobsen discovered uranium in the road cutting near Hawks Crag, the West Coast experienced a short but intense uranium rush, and the hills rang with the sounds of clanging rock hammers and ticking Geiger counters. The prospectors — labourers, businessmen, chemists, miners, and weekend trampers and deerstalkers — were not the only people excited. The uranium finds raised hopes for a return to the prosperous days of the West Coast gold rushes, when packed public houses and dance halls lined the streets and West Coast ports were the busiest in the country. It might have had little arable land, but the West Coast was seen as a region of natural abundance: uranium could be added to the list of existing economic resources of gold, coal and timber.

Uranium was a familiar word in the mid-1950s: it was the fuel for the nuclear reactors that were going to revolutionise electricity generation and change the way people lived. It was the raw material that would take the world, including New Zealand, into the 'atomic age'. Amid all this utopian promise, however, uranium was primarily used as fuel for nuclear weapons, like the ones being tested by the United States and the United Kingdom in the nearby Pacific. Even so, no one suggested that New Zealand might have any moral or philosophical opposition to mining and exporting uranium — the local and national response was entirely positive.

As prospectors flocked to the West Coast, more uranium-rich deposits were reported. By 14 November, there were six uranium finds — three in the Paparoa Ranges, one in Buller Gorge, and two close to Reefton — and a new air of hope and prosperity on the West Coast. 'Health Hazards from Radioactive Materials' read a headline in *The Press*, reporting from the United Nations Atoms for Peace conference in Geneva.[7] The warning seemed to go unheeded: hotels made presents of radioactive rock fragments to parting guests; shop windows attracted customers with displays of uranium-bearing rock; and the Berlins Hotel, where Cassin and Jacobsen had spent the night after their uranium find, had its busiest afternoon's trade since the gold rush. At Hawks Crag — where Cassin and Jacobsen had said their uranium find was — a rich terracotta ore at the roadside attracted passing motorists, many of whom took home samples of the rock. There was a rush on Geiger counters and the second printing of Grange's *Prospecting for Radioactive Minerals in New Zealand* was almost exhausted. The mayor of Westport said the discovery of uranium-bearing rock in the Buller Gorge had caused 'a wave of optimism unknown in the district for more than 50 years'.[8]

According to *The Press*:

There has been an unmistakable air of new liveliness in the main street of Westport during the brilliant weather while the 'uranium boom' was at its peak, and there were more cars in the business area on Monday and yesterday than had been seen for years.

Businessmen say that business has shown a returning briskness already. Travellers have been getting better orders. Hotels were crowded out

Even the local ice-cream factory got into the act, offering 'uranium ice cream' to West Coasters in November 1955. *Grey River Argus*, 23 Nov. 1955, p. 5.

yesterday, and some guests were required to go elsewhere or accept makeshift accommodation . . . Hotel talk is unceasingly on the subject.[9]

While excitement reigned on the West Coast, Wellington-based scientists were continuing to analyse the uranium discovery. The quality was there, but was there enough of it for a uranium mine? For minerals like uranium, which are distributed throughout a rock rather than concentrated in seams (as many precious metals are), costly and extensive processing is required, and for a uranium deposit to be economic it would have to exist in quantities and proportions that offset the cost of extraction. The location of the deposit — its depth, geographic spread and the distance from where it would be used — would also factor.

Meanwhile, a DSIR technical report put a dampener on the initial excitement. On 23 November, only nine days after the first uranium claim was registered, more detailed analysis showed that Cassin and Jacobsen's initial sample had anomalously high levels of uranium; further samples from the same location contained one-quarter to one-hundredth of the first sample's levels. While New Zealand's first uranium find was reported as being 'not commercial', the DSIR report said it gave a valuable lead as to further places and rock types in which to search.[10] When he heard about the report, Cassin declared it was 'too silly for words', given that no boring tests had been carried out. He intriguingly told *The Press* that it appeared people had been asked to 'soft-pedal' about it because of 'international complications'.[11]

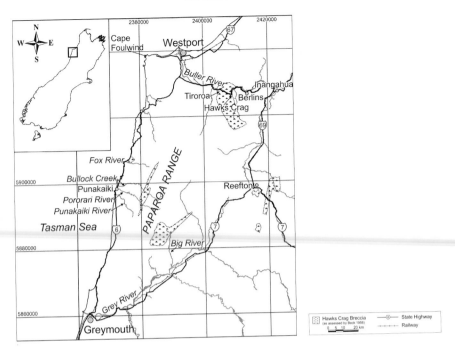

Map showing distribution of the uranium-bearing Hawks Crag Breccia on the West Coast of the South Island of New Zealand. Uranium was found in Buller Gorge, at the Fox River mouth and up Bullock Creek in the Paparoa Ranges. Adapted from Beck *et al.*, 'Uranium Mineralization in the Hawks Crag Breccia of the Lower Buller Gorge Region, South Island, New Zealand', *New Zealand Journal of Geology and Geophysics*, 1(3), 1958, pp. 432–50.

Cassin and Jacobsen's uranium find had initially attracted interest from Australian mining companies — the prospectors told reporters that they had been offered 'unlimited funds' to develop the strike — but following the report that the lode was of no commercial significance the Australian companies withdrew support.[12] Cassin and Jacobsen did no further prospecting of their claim, which irritated the Geological Survey. As Dick Willett, who succeeded Grange as director of the Survey, expressed it, some prospectors were 'content to find a piece of radioactive rock, peg a claim and wait for the State to either present them with a reward or hope for some overseas group to buy them out'.[13]

The prospectors who found Australia's uranium deposits had each received tens of thousands of pounds. Cassin and Jacobsen were not so fortunate. In 1956, they were each awarded the meagre amount of £100 under the Atomic Energy Act. In making the reward, the Minister of Mines noted that 'whilst

In this photograph, probably taken by Tas McKee of Lime and Marble, a helicopter prepares to carry men and equipment to Buller Uranium's prospecting camp on the north side of Buller Gorge in 1956. Author's collection.

it was disappointing that the early reports were not more encouraging as to the commercial value of the area, the discovery was important in renewing interest by prospectors'.[14] Cassin and Jacobsen were 'bitterly disappointed' at the value of the reward — they said that as a result of earlier discussions 'at Ministerial level' they had been expecting at least £10,000.[15]

Prospecting for uranium continued. In May 1956, men employed by Buller Uranium Limited (a subsidiary of the Nelson company Lime and Marble Limited) reported three finds of radioactive boulders and outcrops in the lower Buller Gorge. Uranium Valley, a company set up by two Westport men, found uranium in carbonaceous shales at the Fox River mouth and in the Paparoa Ranges inland from Punakaiki.

Uranium was back in the news. Just a week after these new finds were reported, Professor Gordon Williams, from the University of Otago School of Mines and Metallurgy (he also held a mineral prospecting warrant for a Buller Gorge claim), told a mining conference that uranium 'twice as good' as any ore found previously had been discovered in the Buller region and

In this photograph, probably also by Tas McKee, a helicopter arrives at 'Benney's Landing', site of Buller Uranium's prospecting camp on the north side of Buller Gorge, October 1956. Author's collection.

that there was now 'a distinct possibility of an underground uranium metal mine in New Zealand'.[16]

Unlike the Cassin and Jacobsen find, the 1956 discoveries held up under scrutiny. Over the next four years, the New Zealand Government spent more than £35,000 on the West Coast search for uranium, mostly to fund the prospecting efforts of two companies: Buller Uranium and Uranium Valley. The money from most of the government grants was to be contingently repayable — if the companies ever went into extraction and production of uranium oxide, the money would have to be repaid. Geological Survey and Mines Department staff, who were also working in the area, supported the prospectors. There was no question of the government opposing or not supporting uranium prospecting — uranium was a globally desirable mineral that could boost New Zealand's economy, and was treated the same as any other economic mineral.

In the Buller Gorge names like 'Uranium Creek' and 'Radioactive Creek' started appearing on maps. Buller Uranium used money from their parent company, Lime and Marble, along with their government grants, to cut tracks in the steep bush, make clearings for helicopter airdrops, establish and provision four uranium-prospecting camps, and run a field telephone from the main camp to the trunk line near the Buller River. A government grant was used to cover the cost of a helicopter airlift to transport materials and equipment to a new camp at 'Benney's Landing', named after the Under-Secretary of Mines, 400 metres above sea level on the north side of Buller Gorge. In this area of cleared scrub, prospectors built a sapling platform as a landing stage for a helicopter to fly in material for the camp, and built a hut above the cliff face where the uranium-bearing rock had been found. Prospectors expanded their search, tramping through the rainforest each laden with a geological hammer, a slasher, a compass, a Geiger counter and a scintillometer (an electronic device used to measure gamma radiation). In 1957 and 1958, the Government funded the construction of a jeep track from Tiroroa Siding to the uranium area high above the north side of Buller Gorge.

Uranium Valley built huts at two locations in the Paparoa Ranges, carrying supplies by helicopter, packhorse and on the backs of the prospectors to bases at Bullock Creek and Pororari. In both the Buller and Paparoa locations, rock faces were cleared of vegetation and blasted to expose unweathered rock so that the radioactive seams could be traced more clearly. The work was hard — the conditions were rough, the weather was wet, and measles, influenza, and blood poisoning plagued the prospectors.

Uranium had been in and out of the news throughout the late 1950s, but after the initial burst of uranium fever failed to result in any mineable deposits, public and media enthusiasm had waned. In January 1957, *The Press* report that 'Rich uranium reserves, conservatively estimated at between £10,000,000 and £20,000,000, but probably worth twice as much' had been discovered in Buller Gorge made only a few column inches on page 13 of the newspaper.[17] Ernest Marsden remained characteristically enthusiastic, however, and that June he told the Hutt Valley Chamber of Commerce that within a few years West Coast fields would be producing enough uranium to run a nuclear reactor in New Zealand. Ever optimistic, Marsden saw the issue of nuclear power for New Zealand as being a question of 'when', rather than 'if', predicting that New Zealand would be running its first nuclear reactor in ten years' time.[18]

REWARDS UNDER THE ATOMIC ENERGY ACT

As the number and significance of the uranium finds grew, so did dissatis-
faction with the Government's level of support for uranium prospecting,
and the paltry rewards offered under the Atomic Energy Act. Lime and
Marble Limited, which had put the most money and effort into a systematic
prospecting programme, criticised the lack of information about market
prospects, and said the Atomic Energy Act, which was passed before any
uranium was found in New Zealand, did not meet present conditions.
'Unlike Australia, USA, or Canada,' said Lime and Marble head Tas McKee,
'there is no Government buying agency offering guaranteed prices or mine
development allowances, and bonus payments. Furthermore, there have been
no tax concessions available such as . . . The Australian provisions which . . .
allow complete remission of income tax for all profits from uranium until
1965.' McKee concluded by saying: 'it seems fairly certain there will be no
investments of private overseas capital in New Zealand uranium until terms
as attractive as in Australia are offered'.[19]

Lime and Marble, along with the Mines Department, the DSIR and
the University of Otago, were consulted in framing the Atomic Energy
Amendment Bill, which was passed in October 1957. The Atomic Energy
Amendment Act, modelled largely on the Australian legislation, took away
the confiscatory provisions of the 1945 Act, and gave the owner the right to
sell uranium ore at current market prices, either to the Minister of Mines or,
with the minister's permission, to other parties. The Act also established a
schedule of rewards for discoveries of uranium. Cassin and Jacobsen were
the first to be acknowledged under the new schedule, with payments of
£400 each paid to Charles Jacobsen and to the estate of the late Frederick
Cassin. Further rewards, of up to £1,000, were paid to Lime and Marble,
Uranium Valley, and to Hamilton company director Robert O'Brien, all for
discoveries of uranium in Buller Gorge and the Paparoa Ranges.

THE UKAEA COMES TO NEW ZEALAND

The UKAEA, which had heard about New Zealand's uranium finds from
Marsden, sent their chief geologist to the West Coast in late 1956. He said
the Buller Gorge area showed promise as a uranium mining field, and was

Lime and Marble staff pose outside the adit on the south side of Buller Gorge, in 1960.
The first rock recovered from the hole yielded uranium oxide levels higher than those
found at the UKAEA camp across the river. Courtesy Jock Brathwaite.

'a splendid chance for the New Zealand metal mining industry to get on its feet and to put this country "on the map" as a producer of the newest kind of fuel — the raw material for nuclear energy'.[20]

To assist in the development and exploration of the area, the UKAEA sent a geologist and a mining engineer to work on the West Coast. By now, the Geological Survey had two geologists mapping stratigraphy and structure in Buller Gorge, along with a Wellington-based petrologist working full time on mineralogy. While private companies working the area kept the results of their investigations secret, the government geologists published the results of their research, adding to the body of knowledge about the local geology and the best prospecting methods. Under the guidance of the West Coast's inspector of mines Lloyd Jones, a survey party from the Mines Department was surveying all outcrops and preparing accurate plans of the mineralised portions of the Hawks Crag Breccia to help Buller Uranium in their search. Even so, the Mines Department later said that the UKAEA geologist was the first person to sample the deposit on a systematic basis. As a result of a report by the UKAEA mining engineer, the UKAEA agreed to finance sampling operations in the Buller Gorge, provided they would be given first right of refusal over any uranium found in New Zealand. In response, the *Greymouth Evening Star* reported 'a new prosperity is expected for the West Coast'.[21]

After protracted negotiations between Buller Uranium and the UKAEA, a confidential agreement between the Crown, the UKAEA and Buller Uranium was signed on 11 March 1959. The agreement arranged for the UKAEA to carry out a three-stage investigation of the uranium deposits on the north side of the Buller Gorge, with each stage contingent on success in the previous stage. As part of stage one, they set up an office and laboratory in Westport with Bill Hill as their resident geologist. Unfortunately, however, the lengthy delay in finalising the agreement — caused by some issues Buller Uranium had with the details of the agreement — meant that by the time the agreement was signed the market for uranium was deteriorating. According to the Mines Department, the UKAEA were by now aware that the deposit was unlikely to be economic, on current world markets, but were obliged to fulfil the terms of the contract. *The New Zealand Herald* later reported that the UKAEA 'was persuaded to spend time and manpower on investigating the New Zealand uranium potential at a time when it was already assured of all the supplies it needed for some years ahead, and at a time when some other countries had more substantial claims for their resources to be investigated'.[22] It can be seen as representative of the closeness between the two countries that it was almost seen as a given that New Zealand would offer its uranium supplies first to the United Kingdom — but only if they were not needed domestically.

As part of the fulfilment of the agreement, the UKAEA funded the extension of Buller Uranium's access road on the north side of Buller Gorge, at the end of which prospectors set up a camp with bunkhouse, drilling equipment and a rock crusher. The stage-one drilling programme ran from September 1959 to March 1960. Ten short adits, or tunnels, were drilled by jackhammer, and once a clean rock face was exposed a 2-ton sample of rock was blasted out of the rock face onto collecting tarpaulins mounted on wooden frames. The rock was put into drums then passed through a rock crusher, and sent to the Geological Survey where the samples were broken up for analysis. The results were disappointing. Instead of the strong uranium-rich layer that had been hoped for, there was only a series of small, low-grade, isolated lenses of uranium-rich rock. Further investigations of the area were not warranted, and in August 1960 the UKAEA terminated the contract.

Prospecting, meanwhile, was continuing in other areas. Lime and Marble, with more government funding, set up a new camp and drilled an adit on the south side of Buller Gorge which immediately yielded uranium oxide levels higher than those found in the UKAEA programme north of the river.

URANIUM GLUT

By 1960, the predicted 'atomic age' had not really begun. The United Kingdom and the United States were operating only five electricity-generating nuclear reactors between them, and the primary purpose of the British reactors was the production of weapons-grade plutonium. The United Kingdom stations at Calder Hall and Chapelcross were commissioned in 1956 and 1959; and the United States stations — Shippingport, Yankee Row and Dresden 1 — between 1957 and 1960.

There was a glut of uranium on the world market, and prospects for sale of any New Zealand-sourced uranium oxide were poor. Since New Zealand's search for uranium had begun in 1954, uranium had gone from relative scarcity to abundance on the global market. At the same time, Western production of 43,000 tons per annum — coming mainly from the United States and Canada, with lesser amounts from South and Central Africa, Australia and France — was exceeding demand, most of which still came from the military. Some countries were stockpiling uranium for defence purposes, and some overseas mines had closed. Uranium-producing mines in the West held contracts for delivery of uranium oxide concentrates to the UKAEA and the USAEC under which the prices paid for uranium incorporated amortisation of the mines and treatment plants. When the contracts expired in the 1960s, any new producers of uranium oxide would have to compete with producers with fully amortised plants, who were predicted to be able to drop the price of uranium oxide from NZ$8–10 per pound to as low as $4 a pound. This meant that to make ore grade, any uranium-bearing rock would have to contain at least 0.4 per cent uranium oxide.

In Australia, exploration for and development of uranium prospects was virtually at a standstill, and the Australian Bureau of Mineral Resources advised New Zealand not to consider establishing a uranium mining and treatment industry until there was evidence of firm contracts at a suitable price. In the previous five years, the New Zealand Government had spent £36,902 on financial assistance to uranium prospectors (with a further £17,775 contributed by the UKAEA) and the latest prospecting results did not warrant further expenditure. In March 1960, five months before the UKAEA terminated their contract, the Mines Department had recommended that no further government money be spent on uranium prospecting.

West Coasters, along with prospectors such as Tas McKee of Lime and Marble, were scornful of the Government's decision. Gordon Williams sided with the angry locals, and said that the newly explored south side of the Buller Gorge was 'a most exciting prospect by any standard'.[23] In a letter to Williams, the Minister of Mines attempted to put him in his place, by saying that 'uranium has in New Zealand become invested with rather mystic attributes and, as you yourself have mentioned, publicity in the press has contributed to this situation. When all is said and done the assessment of a uranium prospect must be based on the same fundamental mining engineering principles as are applied to any other mineral deposit and this seems on occasion to have been lost sight of.'[24]

Buller Uranium and Uranium Valley responded to the Government's decision by complaining, in person, to the Minister of Mines, who asked for a written submission. Lime and Marble and Uranium Valley responded by jointly preparing a proposal for a comprehensive exploration and prospecting plan for the Buller–Paparoa uranium province. The plan included expanding consultant and laboratory services, drilling, radiometric surveys, geological mapping, mineralogical and petrological work, and field prospecting, and was expected to cost around £250,000. They said that 'if the programme suggested is adopted it will be in line with what has been carried through in other Commonwealth countries, where Government actively encourages prospecting'.[25] But when Cabinet met on 20 June 1960, on the recommendation of the Minister of Mines, the proposal was declined.

RENEWED INTEREST IN WEST COAST URANIUM

Lime and Marble retained their interest in uranium prospecting, suggesting first that New Zealand invite the International Atomic Energy Agency to

New Zealand to assess its uranium deposits, and then that New Zealand approach EURATOM, the European Atomic Energy Community, with a proposal to carry out a proving programme. Then, in 1964, when the Electricity Department projected that New Zealand would need its first nuclear power station in the 1970s, Lime and Marble resumed their push to prove Buller Gorge uranium deposit.

Since the late 1950s programme of exploration, some New Zealand officials, including staff at the Mines Department, had suggested it would be possible for New Zealand to manufacture fuel for a domestic nuclear reactor by processing uranium-bearing ores to uranium oxide in New Zealand. The Electricity Department liked this idea, as by 1966 they were estimating that a future nuclear power station would have to spend £5–6 million a year on imported uranium fuel. And, based on known reserves, a world shortage of uranium was predicted in the 1970s. If new deposits of the present minimum grade were not found, then lower-grade or less extensive deposits — like those known to exist in Buller Gorge — might become commercially viable to work. Globally, there was increasing optimism in the uranium industry, and the search for new uranium deposits was intensifying. As forecasts of future nuclear power capacity were revised upwards, so was the predicted demand for uranium. By 1967, projected growth in the global nuclear electricity-generating sector meant that 500,000 tons of uranium needed to be found by 1981. As the result of projected demand, global exploration was at an all-time high.

The drive to resume investigations of New Zealand's uranium deposits intensified. In February 1966, the New Zealand Atomic Energy Committee recommended a full appraisal of New Zealand's uranium deposits. In August, a party of six of Australia's leading nuclear scientists visited New Zealand for a week. The leader of the group got the attention of the media by predicting that world price for uranium oxide could reach US$30 a pound within ten years, which would make West Coast deposits economic, and by saying that it might be possible for New Zealand to supply Australia with heavy water made at Wairakei and for Australia to supply New Zealand with uranium. He also said that 'both Australia and New Zealand were on the threshold of nuclear power and a lot of expertise had to be developed.'[26] This made the front page of *The Press* — someone from overseas proclaiming that New Zealand had economic deposits of uranium was taken a lot more seriously than a New Zealander making the same claim. In October, the Geological

Survey completed a proposal to recommence and extend the investigation into New Zealand's uranium deposits.

In mid-1966, Uranium Valley had applied to renew their mineral prospecting warrant over the Bullock Creek–Pororari River area, leading the Mines Department to review its policy with regard to uranium prospecting. While previous prospectors had not discovered any economic uranium deposits on the West Coast, more work needed to be done to conclude whether no deposits existed or whether there might be lower-grade deposits, or high-grade deposits of smaller extent, that might become economic if uranium prices rose. Uranium Valley's application was for the one area where, in the opinion of the Mines Department, more work was needed; the uranium deposits were not well understood, and more advanced prospecting — including drilling — was required.

The Mines Department, however, knew that Uranium Valley did not have the financial and technical resources for such a programme, so recommended to the Mineral Resources Committee that uranium prospecting officially recommence, with the first stage being financial support for Uranium Valley's investigation of the siltstone beds of the Hawks Crag Breccia exposed in the Pororari River. Other money was made available for additional reconnaissance work, and for more detailed investigations of South Buller Gorge. Financial assistance to mining operations, in the form of loans and grants, was part of standard Mines Department practice, but the level of support for uranium prospecting, as compared to other mining operations, was significant.

The following year, on the invitation of Lime and Marble (which was now a 50 per cent shareholder in Uranium Valley as well as parent company of Buller Uranium), Australian company CRA Exploration purchased an option to look for new uranium locations in the Buller Gorge, Pororari River and the Fox rivermouth areas. Between April and September 1967, CRA spent A$37,299 on exploration, including an extensive helicopter scintillometer survey to measure radiation intensity. The survey found no new uranium prospects — rather it confirmed that the main mineralised areas had been located by previous ground surveys. CRA determined that mining these areas would not be economical.

In late 1968, Lime and Marble used government grants to drive two adits at Uranium Valley's Paparoa sites and to drive a 300-foot (about 90-metre) adit on a claim held by Buller Uranium in Buller Gorge. By now, the health effects of working with radioactive materials were more widely known, and the National Radiation Laboratory[*] provided radiation monitoring badges for prospectors working in the adits. Drillers were alerted to 'the necessity for good personal hygiene both bodily and with respect to clothing', and — in the days before cigarette smoking was conclusively linked to lung cancer — to avoid the cancer risk of inhaling radioactive rock particles with their cigarettes they were told that 'hand rolled cigarette smoking' should be 'limited to cigarettes rolled before commencement of a shift'.[27]

While the tunnelling did not reveal any economic deposits in the Pororari area, the adits driven in the until-now neglected South Buller area indicated ore-grade material that warranted further investigation. Buller Uranium proposed a South Buller drilling programme at a cost of $65,280. The Mines Department supported the proposal, and Cabinet approved a grant of up to $41,640 for Buller Uranium, at the rate of one government dollar for every dollar the company spent. Because this new subsidy rate (previous grants had been at 4:1) left a greater financial burden on the company, they sought outside help and reached an agreement for a joint venture whereby Carpentaria Exploration Co. (an Australian company owned by Mt Isa Mines) would fund and supervise the drilling in return for an option over mining rights. Drilling started in early 1971, and two shifts of drillers operated a diamond drill 24 hours a day for nearly two months. Stage one of the programme was completed by August, when drill core assays were sent to Australia for analysis. Once again, the results from the drilling were not encouraging and a planned second stage did not proceed.

GROWING INTERNATIONAL INTEREST

Investigations until now had tended to progress when global demand for uranium was high and come to a standstill when demand was low, or, in the case of international companies, when they had better prospects in other countries. As a result, the West Coast's uranium had still not been proven — no one knew with any certainty the full geographical extent or concentration of New Zealand's uranium deposits.

---

[*] The Dominion X-ray and Radium Laboratory changed its name to the National Radiation Laboratory in 1963.

The latest predicted shortfall of recoverable uranium in the Western world — of one million tonnes by the year 2000 — meant that substantial new uranium deposits needed to be under development by the mid-1980s. West Coasters remained optimistic about their uranium reserves. Aware of the new demand projections, in June 1972 the Labour MP for Westland, Paddy Blanchfield, asked Leslie Gandar, Minister of Fuel and Power in the National Government, 'what preparations have been made to utilise the uranium deposit in the Buller Gorge?' Gandar replied that the Government had no proposals to recover uranium from Buller Gorge, reiterating that the region's uranium concentrations were too low to be considered an economic source of uranium.[28]

Some international companies disagreed. Companies from Italy, the United States and Germany contacted New Zealand officials to express their interest in prospecting for uranium in New Zealand. When they investigated more closely, however, they discovered that the most promising uranium provinces were already held by Lime and Marble and its subsidiaries. In April 1973 Uranerzbergbau-GmbH & Co. KG (Uranerz), a government-financed company searching for uranium supplies for West Germany, approached Lime and Marble. In October that year, representatives of both companies met with the New Zealand Government, now led by Labour's Norman Kirk. Following negotiations with Lime and Marble, Uranerz was prepared to prospect areas in Buller Gorge and the Paparoa Ranges.

In their first year in New Zealand, Uranerz planned a substantial $150,000 prospecting programme, which would include aerial surveys and surface exploration. In the drafted joint-venture agreement, Uranerz would have financial responsibility for the venture, and Lime and Marble would be given the option to take up to 40 per cent of the equity capital when the uranium was mined. The plan was to mine the ore and treat it close to the site, exporting the extracted uranium oxide.

The Government approved the joint venture in 1974, subject to New Zealand equity in any mining operation being set at a minimum of 50 per cent, and the negotiation of suitable terms for export of the uranium oxide. Negotiations continued throughout 1975, and through a transition from a Labour to a National Government, with particular emphasis on the amount of uranium oxide that would be available for export. At this time, serious consideration was being given to the need for nuclear power to contribute to New Zealand's growing electricity demand; the discovery of a supply of uranium in New Zealand would have an impact on the choice of nuclear reactor. Based on current demand projections and the installation of a natural uranium-fuelled reactor, the Commissioner of Energy Resources recommended that in the event of a major uranium discovery, a supply of at least 5000 tonnes of natural uranium* should be reserved for local use. In the longer term, the New Zealand Electricity Department estimated that 20,000 tonnes of uranium would be needed to meet the 30-year lifetime needs of thermal reactors planned to come on line before 2000, with the expectation that fast-breeder reactors — which produce more fissile material than they consume — would be introduced in 2000.

After further negotiations with Uranerz, it was proposed that they could export 25 per cent of the first 10,000 tonnes of uranium concentrate produced, and 50 per cent thereafter. In March 1977 Uranerz agreed to the export formula, provided the New Zealand Government purchase uranium from Uranerz at world-market prices. New Zealand Treasury, aware that the price of uranium had been very unstable with recent steep price rises, was concerned that 'New Zealand could be faced with a rapidly escalating price which it would have to pay for an indigenous resource — and which would not bear any relationship to costs of production in New Zealand'.[29] Concerns were also raised over New Zealand's obligations under the Nuclear Non-Proliferation Treaty.

At the same time, Australia was grappling with the issue of uranium mining. In September 1977 the Australian Council of Trade Unions, led by Bob Hawke, demanded a referendum to decide whether uranium mining should go ahead at the Ranger Mine in Kakadu National Park, but the Liberal–National Coalition Government rejected their demand. The Australian Labor Party was also voicing dissent. In a televised statement on 4 September 1977, Labor Party leader Gough Whitlam attacked the Government's decision, saying that the course Australia was following would add to poisonous nuclear wastes for which no safe disposal and storage technology had been devised, and could add to the risk of nuclear war.

Back in New Zealand, negotiations between the New Zealand Government and Uranerz continued, and agreement on an export formula was

---

* Uranium occurs naturally as uranium oxide. Processing can extract the natural uranium, with about 1.2 tonnes of uranium oxide needed to produce 1 tonne of natural uranium. Enriched uranium is natural uranium that has been processed to increase the proportion of fissile uranium-235, a naturally occurring isotope of uranium that is usually only present at less than 1 per cent.

finally reached in early 1978. Uranerz began their on-ground investigation programme in March. After five years of negotiating, Uranerz's involvement with the West Coast uranium was short-lived: the results of two months' fieldwork were disappointing, and they surrendered their prospecting licence in 1979.

Uranerz's investigations in the late 1970s turned out to be the last focused investigation for uranium in New Zealand. After 25 years and hundreds of thousands of dollars spent looking for and attempting to prove the West Coast's uranium resources, no economic deposits were ever found. By the time that Uranerz abandoned their search for uranium, the mineral was no longer considered important for New Zealand — a substantial natural gas field had been discovered in offshore Taranaki, and nuclear power stations had been deleted from the New Zealand power plan in favour of gas turbines.

While concern about bomb testing emerged in the 1950s, and led to protests in the 1960s and 1970s, this never extended to efforts over this same time period to establish an industry mining uranium, which — as well as being fuel for nuclear energy generation — was used in nuclear weapons. It would not be until after the introduction of New Zealand's nuclear-free legislation that mining radioactive minerals would be seen in a sinister light.

5

★

There's strontium-
90 in my milk
Safety and public
exposure to radiation

The early pioneers learned by bitter experience that they were using potentially lethal agents, and many of them paid with their lives for the knowledge they had gained in exploring this new field. But their death was not in vain and their suffering helped to guide those who came after them. — GEORGE ROTH, DIRECTOR, DOMINION X-RAY AND RADIUM LABORATORY, 1952[1]

Most people date their awareness of radiation hazards from the moment they became aware of the danger to themselves from nuclear warfare.
— GEORGE ROTH, DIRECTOR, DOMINION X-RAY AND RADIUM LABORATORY, 1965[2]

My mother, who was born in Christchurch in 1943, says one of the highlights of a trip to town with her own mother was getting to play with the x-ray machine in the shoe shops. These 'pedascopes', promoted as an ingenious way to ensure that your foot fitted correctly inside any given shoe, were just as likely to be used by children as a toy:

It was towards the end of the 1940s when I first recall placing my feet in an x-ray machine in a shoe shop. That was what we did when we tried on new shoes, to see if there was room left in front of our toes for growth. Not only did my mother and the shop assistant look down the viewing tube to check my feet, the tube height was adjustable so children could get it low enough to look down themselves. While my mother tried on several pairs of shoes herself, I would be 'playing' with my feet in the x-ray machine. I had such fun; I can still see those little metatarsi going in, out, in again. I didn't ever want to leave the shop. To see parts of your own skeleton was a major childhood event, and it was a sad day for children's psyches when such a fun thing had to stop.[3]

She was nine years old before New Zealand's Radiological Advisory Council issued shoe shops with a warning notice that advised that pedascopes could be hazardous and imposed restrictions on their use. Even so, the last shoe-fitting x-ray machines were not removed from New Zealand shoe shops until 1969.

While the end of the Second World War and the nuclear arms race that followed brought a new type of radiation to come to terms with — radioactive fallout from nuclear weapons — attitudes to other forms of radiation were slow to change. There was a clear separation between nuclear weapons — seen as abhorrent by some, and as a necessary evil by others — and what were considered 'peaceful' uses of the atom. Although, as physicist George Roth said in 1965, the awareness of the danger to humans from nuclear warfare and nuclear bomb testing did eventually lead to a greater awareness of the dangers to humans of all radiation hazards.

RADIATION PROTECTION LEGISLATION

As covered in Chapter One, in the 1920s the Health Department took the initial lead in warning New Zealand hospitals and medical practitioners about the risks of exposure to radiation. For more than 20 years, this advice took the form of alerting medics to international standards and recommendations for radiation safety. After the end of the Second World War, when radioactive isotopes from American cyclotrons became available to New Zealand scientists and doctors, the DSIR became involved, with a Tracer Elements Committee providing advice on health protection for workers handling radioisotopes. The international guidelines they circulated required: workers to wear film badges and have regular blood tests; laboratories to be

A meeting of the Radiological Advisory Council at the Dominion X-ray and Radium Laboratory, Christchurch offices, c. 1955. From left to right, council members were: Dr Peter Allen; Dr Herbert Purves; Bert Yeabsley, deputy director of the laboratory; T. A. Ward, secretary to the council; Harold Turbott, Deputy Director-General of Health and chairman of the council; George Roth, director of the laboratory; Dr Noel Hill; and Athol Rafter from the DSIR. Courtesy National Radiation Laboratory.

checked weekly for radiation contamination; and radioactive material and waste to be safely stored and disposed of.

New Zealand introduced its own radiation protection measures in the 1940s, with the passing of two pieces of legislation initiated after hospital radiologists raised concerns about safety standards. The Electrical Wiring (X-Ray) Regulations 1944 required the registration of all x-ray plants and set out technical requirements for the safety of x-ray equipment. But it was the next piece of legislation, the Radioactive Substances Act 1949, that introduced controls on the use of ionising radiation: anyone who wanted to use, own or sell radioactive substances or irradiating apparatus would now need a licence. The Act also set up a Radiological Advisory Council to provide the Minister of Health with specialist advice on all matters concerning radiation. The council consisted of three radiologists, a senior member of the DSIR, a nominee of the University of New Zealand, a health physicist, and the Director-General of Health (or his deputy) as chairman.

The first set of regulations issued under the Act, The Radiation Protection Regulations 1951, included no specific code of practice, but said that no person should be subjected to a maximum permissible exposure (except in the case of a patient being exposed for medical reasons) and made the licence holder responsible for any breach of the regulations. While the regulations were not prescriptive, a detailed set of recommendations — Recommendations for Protection from Radiation Hazards — were issued in conjunction with them and distributed to hospitals around New Zealand. This publication described x-rays as 'potentially lethal agents' that 'should be treated accordingly' and gave recommendations on how to set up rooms containing irradiating apparatus, and how to protect medical staff from x-rays and radioactive substances.[4]

The Radiological Advisory Council was proactive in promoting safety, and in 1952 it issued a survey of radiation hazards in New Zealand. The publication, written by George Roth (Director of the Dominion X-Ray and Radium Laboratory, and a member of the council), informed practitioners about the type of hazards associated with working with radiation. Roth pointed out that, although there were 650 x-ray plants in use in New Zealand, there were only 45 qualified radiologists or radiotherapists, meaning that most x-ray plants were operated by people 'who cannot be expected to have expert knowledge of the potential dangers of X-ray work and of the methods which make it safe to use ionising radiations'. While acknowledging that most radiation workers were conscious of the need for protection, he said there were still some radiation workers who 'either believe that they are personally immune from the injurious effects of excessive exposure to ionizing radiation or even profess to deride the idea of danger altogether'.[5] Roth had previously written about two general practitioners and one radiologist who had died as the result of radiation injuries, as well as about a dozen cases of 'self-inflicted radiation injuries of various degrees of severity' in x-ray workers, including radiologists, radiographers, medical practitioners, the matron of a small hospital, a chiropractor, a dentist and a university student.[6] Two decades of gentle advice had clearly not worked, but there were still no prescriptive requirements — the new legislation raised awareness and set standards, but it was up to the hospitals and medical clinics how they chose to meet these standards.

DOMINION X-RAY AND RADIUM LABORATORY

After the Radioactive Substances Act was passed, the Department of Health took over responsibility for the Dominion X-Ray and Radium Laboratory

Norman Kirk, then mayor of Kaiapoi, and other visitors outside the new offices of the Dominion X-Ray and Radium Laboratory in Christchurch. Courtesy National Radiation Laboratory.

from the British Empire Cancer Campaign. With the expansion of its responsibilities, the laboratory staff increased from seven to 22 people by 1953. Jim McCahon, one of the physicists involved in Marsden's uranium surveys, was appointed physicist in charge of the radioactive substances section. Following his work making and operating Geiger counters for the uranium surveys, McCahon had gained an MSc in physics from the University of Otago. He had then been seconded to the Atomic Energy Research Establishment at Harwell, where he had gained experience in radiation instrumentation and the measurement of radioactive materials.

The Dominion X-Ray and Radium Laboratory worked closely with radiation workers in hospitals and private practice, taking the approach that radiation safety 'cannot be achieved by laws and regulations alone: it can only come about through the intelligent co-operation of the radiation workers themselves'.[7] Laboratory staff monitored and controlled the importation of all radioactive substances into New Zealand, and each year processed about 20,000 radiation test films — which all radiation workers were now required to wear. The laboratory's diagnostic section looked after radiation protection at the country's x-ray plants (of which there were more than 1000), with staff visiting each installation at least once every three years to measure the level of scattered radiation, check protective barriers, make output measurements, and make recommendations for improvement in practice and technology.

Staff from the therapy section made regular visits to the country's x-ray therapy instalments to calibrate equipment and advise on problems in radiophysics. The radioactive substances section inspected the premises of anyone wanting a licence to import and use radioactive material, was responsible for the safe use of radioactive substances in clinical, industrial and research work in New Zealand, and supplied all the radon used in New Zealand from its radon extraction plant. As Andrew McEwan describes in his history of the laboratory, the remote-controlled radon plant trapped radon gases emanating from a radium solution and compressed it into tiny gold tubes. During the 1950s, the laboratory fulfilled 60 to 90 orders per year for radon supplied for medical use in needles, seeds, phials of ointment, and special applicators.

X-RAYS AND MEDICAL EXPOSURE TO RADIATION

Despite growing public concern about exposure to radioactive fallout from bomb tests, New Zealanders still received a much higher dose of artificial radiation from x-rays than from fallout. As the potential dangers of radiation exposure came to light, the use of x-rays came under closer scrutiny, particularly the non-medical and unnecessary use of x-rays, such as in shoe shops.

By 1957 New Zealand had 60 radiologists, 1084 x-ray plants, 1089 persons licensed to use x-ray plants for specified purposes, and 311 people licensed to use radioactive substances in specified amounts. A decade earlier, New Zealand had had no radiation protection legislation, but Turbott boasted that New Zealand's radiation protection legislation was now 'recognised as the most effective within the Commonwealth, if not the world'.[8] He was right in saying that New Zealand was a world leader — in 1958, only Sweden and New Zealand had a full record of diagnostic x-ray installations.

NUMBER OF REGISTERED X-RAY PLANTS IN NEW ZEALAND ON 31 MARCH 1957

PURPOSE FOR WHICH X-RAY PLANTS ARE USED	OWNERSHIP		TOTAL
	PUBLIC	PRIVATE	
Radiographic or fluoroscopic	219	95	314
Therapeutic	25	28	53
Dental	52	486	538
Chiropractic and naturopathic (diagnostic)	–	51	51
Shoe fitting	–	78	78
Veterinary	4	19	23
Industrial	6	1	7
Miscellaneous purposes (educational, research, testing, demonstrations, etc.)	20	–	20
TOTAL	326	758	1084

Source: Report to UN Scientific Committee on the Effects of Atomic Radiation, 17 Feb. 1958, H1, 26758, 108/11, 1957–58, Archives New Zealand, Wellington, New Zealand.

While radiation workers had been the focus of safety concerns about x-rays, there was now a growing awareness of the potential risk to people receiving diagnostic or therapeutic x-rays. The total radiation dose received from a diagnostic x-ray had dropped dramatically since x-rays were first used, but there were still x-rays being taken which had dubious benefit. In an x-ray screening programme introduced in the middle of the century in an effort to curb the tuberculosis epidemic, hundreds of thousands of healthy people were x-rayed in an effort to find new cases of tuberculosis. In 1941, more than 2000 Wellington factory workers and secondary school students were screened using mass miniature radiography (MMR), a new, cheap form of x-ray photography in which up to a hundred 1-inch chest x-ray pictures could be taken every hour. This Wellington screening programme revealed an active tuberculosis rate of 0.6 per cent.

The first mobile x-ray unit, a local initiative, was introduced in Taranaki in 1946, with a particular focus on screening rural Maori who had a very high rate of tuberculosis. In 1954, the Health Department took over the service, along with other local services, and expanded the scheme nationwide. By 1957 there were nine miniature x-ray units travelling around New Zealand by caravan, taking an annual total of nearly 250,000 miniature x-ray pictures, from which 383 new cases of tuberculosis were identified. That same year, the question of mass chest x-rays as a method of finding new

In the 1950s and 1960s, New Zealanders were encouraged to have an annual chest x-ray to screen for tuberculosis, with mobile mass miniature radiography units travelling to workplaces and marae to make screening easier. Department of Health poster, 1950s, AAFB 24233, box 40/40d, A78, Archives New Zealand, Wellington, New Zealand.

tuberculosis cases was reviewed by the College of Radiologists. The review committee stated: 'In the case of routine X-ray examinations of the chest, the small dose involved is justified provided that a significant number of new cases is being discovered'.[9] Tuberculosis was a deadly disease, and any health risk posed by exposure to radiation as a result of the diagnostic x-rays was considered justifiable.

While New Zealand now boasted the world's lowest tuberculosis death rate, this achievement was credited in part to the MMR campaign. The campaign continued, with its goal the total eradication of tuberculosis. Increasingly, it was targeting what were considered to be the most at-risk groups, such as Maori, whose rates of tuberculosis were up to ten times higher than those of non-Maori. In 1969, the Department of Health advised medical officers of health that MMR programmes be concentrated on 'high-risk' groups — identified as including Pacific Islanders, Maori, freezing workers and psychiatric patients — with every town with a population of 100,000 or more being visited at least every six months.[10] The x-rays were voluntary, and all adults were welcome at the mobile x-ray screening units. A 1972 issue of *Health*, the magazine of the Department of Health, cheerfully advised that 'all persons over the age of 15 years have an annual X-ray'.[11] By this time, the 370,370 people x-rayed over the previous year had resulted in 102 active cases of the disease being found. While x-rays with no medical benefit — like the pedascopes in shoe shops — were now banned, any population-wide health risk from continuing the MMR campaign was outweighed by the reduction in tuberculosis it was believed to lead to. The campaign continued in New Zealand until 1976, by which time the economic and medical benefits of the scheme were in question and the campaign was scaled back.

Genetic studies had revealed the particular vulnerability of the developing cells of children and fetuses to radiation damage, but in the 1950s the risks of radiation were still not fully understood or quantified. One use of medical x-rays that was becoming controversial was x-rays of pregnant women, which although sometimes used in New Zealand were never standard practice. In 1957, the Australian Atomic Energy Commission had advised the Australian Federal Government that 'the exposure of pregnant women to x-rays may incline the child in later life to the disease of leukaemia'.[12] In June that year, a decision of the board of the Sydney Women's Hospital meant that expectant mothers were no longer receiving routine x-rays. In New Zealand, while fetal x-rays were never routine, well into the 1970s it was considered safe for pregnant women to have general x-rays. In 1971 the Department of Health responded to queries from MMR staff about x-raying pregnant women by saying 'there is no risk to the unborn child at any stage in pregnancy during the taking of an X-ray on a MMR Unit. The only radiation involved would be scattered radiation and it is doubtful whether this would even reach the uterus.'[13]

However, children had long been recognised as being vulnerable, with the Department of Health in 1958 recommending against the routine mass x-raying of children. The main stated concern about children being x-rayed was radiation received to the gonad area. Another major childhood source of radiation exposure was from dental x-rays, and in 1957 the Radiological Advisory Council met with representatives of the Dental Association of New Zealand to take steps to ensure that 'radiation received by school children from dental X-rays was reduced to a minimum' and 'no possible genetic

Shoe-fitting x-ray machines, or pedascopes, were promoted as the scientific way of ensuring your shoe was a perfect fit for your foot. Pedascopes were common in New Zealand shoe shops in the 1940s and 1950s, with some still operational until 1969.

damage to children resulted'.[14] Parents, the Radiological Advisory Council said in a statement to the public, 'need have no fear of allowing their children to undergo dental X-rays'.[15]

The shoe-fitting x-ray machines so beloved by my mother and others of her generation had already come under scrutiny. In 1952, when Roth published his survey of radiation hazards in New Zealand, he said that 'excessive exposure to radiation may affect the growth of bones and thus interfere with the normal foot development of children who are fluoroscoped repeatedly when being fitted with new shoes'.[16] That same year, the Radiological Advisory Council sent a warning notice to all owners of shoe-fitting x-ray plants, advising them of the hazards associated with these machines. During a 1954 visit to New Zealand, the American Nobel Prize-winning chemist and nuclear critic Linus Pauling called shoe-fitting x-ray plants 'really terrible, a crime'.[17] Nonetheless, their use continued — a 1957 survey counted 78 shoe-fitting x-ray machines in New Zealand. Controls were introduced, however, and in February 1957 Roth reported that '[a]ll shoe-fitting plants have been reduced in their output to less than about 15 r/min at the level of the foot. They have been equipped with time switches which limit the exposure to about 7 seconds and make it impossible to give a further exposure during the next 35–40 seconds; overlapping lead-rubber aprons have been fitted over the aperture into which the customer puts his foot, and warning notices have been affixed to the shoe-fitting x-ray plants, prohibiting their use by children.'[18] Through the rest of the 1950s and into the 1960s, shoe-fitting x-ray machines were kept under regular surveillance, and by 1964 there were only sixteen machines still in use. The last machines were removed from shoe shops in 1969.

At the same time that unnecessary uses of x-rays were being discouraged, staff of the Dominion X-Ray and Radiation Laboratory's diagnostic section were working with dentists and radiographers to reduce the levels of radiation received by staff and patients. According to Andrew McEwan, the introduction of faster films, added filtration, and use of higher-voltage machines all combined to reduce patient radiation doses to as little as one-twentieth of the original exposure.[19]

MONITORING RADIOACTIVE FALLOUT

While the highest contribution to New Zealanders' annual radiation exposure came from background radiation, followed by medical exposures, there was increasing concern about the small amount of public radiation exposure caused by radioactive fallout. While there were medical benefits to diagnostic x-rays, and background radiation was unavoidable, there was no conceivable good side to fallout associated with bomb testing. Concern about fallout was increasingly widespread and was not limited to the emerging peace movement. Even though many of the fission products deposited in New Zealand and around the world had short half-lives and soon decayed to harmless non-radioactive isotopes, other radionuclides* — such as isotopes of caesium and strontium — were also produced in high yields in the explosions, and had long enough lives to persist in the environment.

Concerns about radiation risks got so high that workers concerned about their own safety took industrial action. In 1956, waterside workers in Wellington declined to unload a Japanese vessel until it had been tested

---

* The word 'radionuclide' refers to a radioactive nucleus with a defined number of protons and neutrons. Carbon-14, for example, has six protons and eight neutrons and is a 'radionuclide'. Carbon-14 is one of three isotopes of carbon (the others are carbon-12 and carbon-13). As carbon-14 is radioactive, it is also known as a 'radioisotope'.

for radiation. (The test results were normal.) In 1957, the Otahuhu Branch of the Amalgamated Society of Railway Servants passed a resolution asking that all fish landed from commercial craft be tested for radioactivity. In 1961, the Auckland Combined Waterfront Union passed a resolution requesting the Department of Health ensure that suitable equipment be provided in Auckland to conduct immediate tests on any radioactive material coming into the Port of Auckland, and that a competent person be available to conduct tests.

Initial fallout measurements in New Zealand were instigated by individual scientists and inspired by scientific curiosity, but, as concerns grew about the impact of radioactive fallout products on human health, more widespread and systematic surveys emerged.

IMPORTANT FALLOUT RADIOISOTOPES AND THEIR HALF-LIVES

RADIOISOTOPE	HALF-LIFE
Strontium-90 (^{90}Sr)	28 years
Strontium-89 (^{89}Sr)	52 days
Caesium-137 (^{137}Cs)	30 years
Iodine-131 (^{131}I)	8 days
Barium-140 (^{140}Ba)	12.8 days

New Zealand was involved in several international programmes to monitor radioactive fallout, some of them in co-operation with the countries responsible for the bomb testing. From 1953 to 1966, as part of a project co-operating with the USAEC, the United States Department of Agriculture measured strontium-90 levels in soils from several New Zealand locations, providing a record of fallout levels in New Zealand. Staff from the DSIR Soil Bureau took the soil samples and forwarded them to the United States for analysis. This monitoring was part of the wider USAEC Project Gabriel, a secret survey to evaluate the radioactive hazards from the fallout of debris from nuclear weapons detonated in warfare. The survey looked at the distribution of strontium-90 by monitoring soil, air and water at about 150 American and international stations. In 1955 the USAEC released a report on radioactive fallout hazards, identifying the main radioactive hazard as strontium-90, but claiming that there was no cause for concern from existing fallout levels. The following year, the British Himsworth Report highlighted the hazards to human beings of ionising radiation from *all* sources — natural, medical, industrial and military — but it too carried a warning about strontium-90: rather than saying there was no cause for concern, it carried an alert that levels of strontium-90 could become harmful if nuclear bomb tests continued.

As part of a worldwide monitoring programme, the Environmental Measurements Laboratory of the United States Department of Energy began monthly measurements of strontium-90 in rainwater at Wellington in 1959. The USAEC also contracted the DSIR's newly established Institute of Nuclear Sciences to measure concentrations of various fission products — including strontium-90, caesium-137 and barium-140 — in Wellington rainwater. This monitoring took place from 1959 to 1970, after which the Institute of Nuclear Sciences continued its own measurements until 1984, by which time levels of these radioactive isotopes had become undetectable. As part of its worldwide fallout monitoring programme, the UKAEA monitored strontium-90 in Ohakea rainwater from 1955 to 1965, with caesium-137 monitored after that.

Other monitoring programmes were locally initiated. As part of the International Geophysical Year programme in 1957, the DSIR set up stations in and about New Zealand to collect rainwater and determine its radioactivity. Monthly rainwater collections were sent to the Dominion Physical Laboratory for evaporation, processing and counting of radioactivity levels. At the end of 1958, the Dominion X-Ray and Radium Laboratory took responsibility for the DSIR network of eight monitoring stations — in New Zealand (Auckland, Wellington, Havelock North, Greymouth, Christchurch and Invercargill), Fiji and Campbell Island — which they used to continue measuring levels of strontium isotopes and total beta radioactivity. Sites were added at Kaitaia, New Plymouth and Dunedin in 1962. Monitoring by the now-renamed National Radiation Laboratory continued until 1985, by which time detectable levels were very low, and monitoring was reduced to the stations at Kaitaia, Hokitika and Rarotonga.

In July 1957, Willard Libby, the American scientist who pioneered radiocarbon dating, wrote to Athol Rafter — director of the DSIR Division of Nuclear Sciences, and New Zealand's own radiocarbon dating pioneer — to ask if Rafter would collect samples associated with Project Sunshine. This secret study of worldwide radioactive fallout patterns involved the analysis of radionuclides in samples of soil, plants and animals, and, in some cases (although not, so far as I have discovered, in New Zealand) human bones and teeth. In one gruesome part of the study, the bodies of stillborn babies from several different countries (again, not New Zealand) were cremated without their parents' permission, with the remains tested for total levels of radioactivity.

The USAEC contract with the Division of Nuclear Sciences stated that the aim of the project was 'to study the nature of the precipitation mechanism and the variation of local rates of precipitation with seasons', with results to be given to Libby in order 'to help settle questions of global atmospheric circulation'.[20] It was not publicly announced, but the purpose of the project was to gather information about the transport of radioactive debris about the globe. The Division of Nuclear Sciences was already monitoring several radionuclides, and Libby's request involved little extra effort. The USAEC accepted the Division of Nuclear Sciences' request for support, supplying capital items worth US$12,880. The programme involved collecting rainwater at Gracefield, near Wellington, and after each rain determining the concentration of strontium-90, caesium-137 and barium-140. One of these radionuclides (barium-140) has a half-life of only thirteen days, and so detection was designed to try to show the speed of fallout from recent bomb tests. The contract expired in 1962.

The International Atomic Energy Agency (the IAEA) also funded some fallout monitoring carried out by the Institute of Nuclear Sciences. In 1962, the Institute of Nuclear Sciences applied for funding for a project to investigate the distribution of the radionuclides strontium-90, caesium-137, cerium-144, zirconium-95, promethium-147 and antimony-125 in a series of ocean-water profiles and surface-water stations in the South Pacific. A further aim of the research was to measure the assimilation of some radionuclides by marine organisms. The IAEA funding paid for equipment to the value of US$9,780 and supplies of US$2,660, which were used to measure radioactive elements in the water of the South Pacific, and would allow distribution patterns of fallout from nuclear tests to be studied and correlated with measurements already being made of other radioactive elements, such as carbon-14 and tritium.

Although not part of the research contract, the DSIR noted that the equipment would also allow the institute to study the build-up of radioactive substances, particularly strontium-90, in the teeth of young children. By this time, the Dental Research Institute had an extensive collection of teeth of children living in areas of known soil types. In 1963, Bill Hamilton, head of the DSIR, advised the Minister of Scientific and Industrial Research that it was proposed 'to cooperate with the Dental Research Institute and Soil Bureau, in an update study of this important and dangerous fission product'.[21] The Institute of Nuclear Sciences contract was renewed on 8 April 1964 for another $2,600 for supplies.

Some fears of fallout were vague and generalised, but the way strontium got into the food chain and could find its way into human bones was a source of more concrete fears about human health, particularly in children. New Zealand had long prided itself on producing enough milk, meat and agricultural produce to feed itself and export to Mother England. Milk from the country's dairy herd was promoted as an essential foodstuff for infants, and daily milk had been provided free to kindergarten and primary schoolchildren since 1937. In the 1950s, however, it was revealed that New Zealand's milk — credited with building 'better babies and strong bodies, bones and teeth in growing children'[22] — was also providing New Zealand children with daily doses of radioactive strontium-90 and caesium-137. These new radioisotopes — they did not exist on Earth before the hydrogen bomb tests that started in 1952 — were mistaken by the human body for calcium and stored in bones alongside calcium, where their radioactive decay would provide a constant dose of radiation to the surrounding tissues.

The Dominion X-Ray and Radium Laboratory/National Radiation Laboratory monitored strontium-90 and caesium-137 — and occasionally strontium-89 and iodine-131 — in milk samples from nine New Zealand regions from 1961. Results show that concentration of strontium-90 and caesium-137 in cows' milk peaked in 1965, with the highest levels recorded in any one month being from Westland and Taranaki in 1965. During the peak fallout period of 1963–66, levels of strontium-90 and caesium-137 in New Zealand averaged about 40 per cent of northern hemisphere levels; but in later years, after 1966, levels of caesium-137 were similar to those in the northern hemisphere — meaning that in New Zealand caesium-137 rather than strontium-90 was the fission product responsible for giving humans the biggest radiation dose.

Public awareness of the dangers of strontium-90 increased after November 1959, when Australia's ABC News reported that the New Zealand soil survey had revealed a startling increase in strontium-90 between 1953 and 1958. The next month, the *New Zealand Public Service Journal* outlined how

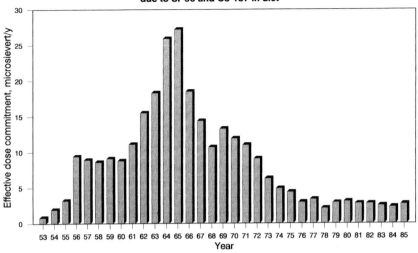

**Annual dose commitments
due to Sr-90 and Cs-137 in diet**

This graph shows the annual dose New Zealanders received from dietary strontium-90 and caesium-137 from 1953 to 1985. The peak levels of strontium and caesium uptake in 1964 and 1965 coincided with the deposition of fallout from the Phase 2 weapons tests by the United States and the Soviet Union. Source: Murray Matthews, *Radioactive Fallout in the South Pacific: A History. Part 3: Strontium-90 and Caesium-137 Deposition in New Zealand and Resulting Contamination of Milk,* National Radiation Laboratory, Christchurch, 1993, p. 106.

strontium-90 arrived in the soil from fallout from bomb tests, travelled from the soil to pasture grass, from the grass to cows' milk, and from milk to lodge in human bone, where it remained radioactive for years and could cause bone tumours, leukaemia and other diseases. Commentaries about the significance of the levels of strontium found in New Zealand varied. The reported results showed that strontium levels at one of the South Island collection sites had trebled between 1956 and 1958. While the acting prime minister, Labour's Jerry Skinner, assured the public there was 'nothing to fear', the director of radiophysics at Wakari Hospital, described the figures as 'alarming'.[23] The Dominion X-Ray and Radium Laboratory, however, continued to assure the public that they were at no risk from radioactive fallout, pointing out that New Zealand's levels of contamination were lower than those in the northern hemisphere, and that the situation was being closely monitored. Strontium levels peaked in 1965. In 1967, at the request of the Education Boards, the Government dismantled the free milk in schools scheme. Coincidentally, this was just two years after the levels of strontium-90 and caesium-137 peaked in New Zealand's milk.

"Charlie and me won't take tea for supper, we'll stick to beer—don't want to take any chances on the milk being radio-active."

This Nevile Lodge cartoon, from July 1966, followed reports of the first French nuclear test in the Pacific. *The Evening Post*, 5 Jul. 1966, p. 14. Courtesy of the Lodge Family.

In response to public concerns, the Dominion X-Ray and Radium Laboratory/National Radiation Laboratory tried to minimise the impact of bomb fallout in relation to natural radiation sources. In 1962, for example, Bert Yeabsley, acting director of the Dominion X-Ray and Radium Laboratory, responded to the National Council of *Parent and Child* magazine's concerns about radioactive fallout by saying 'there is no cause for alarm; indeed when the true facts are realised there is no cause for even mild concern'. He said that the fission product strontium-90 added 'less than 1 per cent to the total radioactivity in our soil'.[24]

The public, however, was becoming more aware of the hazards of radioactive fallout and wanted better access to information. In 1961, after lobbying from the Public Service Association (PSA), the Department of Health began issuing quarterly reports on the levels and health impacts of radioactive fallout in New Zealand. By now, opposition to Pacific nuclear testing was widespread. Britain had stopped testing in 1958, and the United States' last Pacific tests were in 1962, but France began its Pacific testing programme in 1964. The New Zealand Government's official position was now against nuclear testing in the Pacific, and New Zealand scientists were playing an increasing role in providing public information about bomb tests and fallout levels. In November 1965, Rafter boasted that were the French Government to carry out secret atmospheric nuclear testing in the Pacific, New Zealand would know about it within days.

OTHER SOURCES OF RADIATION

In contrast to the widespread public concern about radioactivity from bomb tests, there was a surprising degree of tolerance of some other — also non-beneficial — sources of radioactivity. As we have seen in earlier chapters, one longstanding source of radiation exposure was luminous paint, which was still often used in watches and in some industrial plant. Roth made an

FOR RHEUTMATISM, ARTHRITIS, GOUT & NERVOUS DISORDERS

An early advertisement for a siphon system with a radon bulb — along with carbon dioxide a small amount of radon gas was added to the water. Source: John Campbell, *Rutherford: Scientist Supreme*, AAS Publications, Christchurch, 1999, p. 304. Courtesy John Campbell.

informal survey of Auckland watchmakers in 1949 to investigate the use of radioactive luminous paint in repairing luminous watch dials and hands. It was his observation that watchmakers were not aware of the dangers, were taking no precautions, or were secretive about having radioactive paint, despite non-radioactive luminous paints becoming available after the Second World War. Roth estimated that the average New Zealand watchmaker spent five to ten minutes per week engaged in radium dial painting, but said that no cases of radium poisoning had yet been recorded. He concluded that there was 'no significant hazard to watchmakers in New Zealand through the use of radium activated luminous paint either from the actual luminising itself or from the repairing of watches and clocks with radioactive luminous paint'.[25]

Some people, however, were still willing to subject themselves to unnecessary radiation in the mistaken belief that it was somehow beneficial; despite all of the publicity to the contrary, there were still people who believed in the health-giving benefits of radiation. In 1954 — more than 20 years after radon water and other radium-based 'health' products had been shown to be harmful and potentially deadly — a Wellington firm, Messrs Claude W. Batten and Co., were advertising Radon Sparklets Bulbs, a product to add to their syphon system to make radioactive water. The advertisement for Radon Sparklets Bulbs stated that:

> Radioactive water, consumed as a table water, is of considerable value in increasing the vitality and healthy state of the system. It is an **all-round tonic** and invaluable as a treatment for **rheumatism, arthritis and nervous disorders**. By using RADON SPARKLETS BULBS with a Sparklets Syphon, Radioactive Water can be prepared at home. Each bulb contains a fixed quantity of the essential radioactive element which is the source of Radon (Radium Emanation), so that benefit normally obtained by residence at a Spa can now be conveniently and inexpensively procured without interference with the ordinary routine of daily life.[26]

The Dominion X-Ray and Radium Laboratory tested samples of Radon Sparklets Bulbs and found they contained 0.5–10 micrograms of radium. This made the bulbs 'radioactive substances' under the Radioactive Substances Act 1949, which prohibited the manufacture, sale, import or export of any radioactive substance without the consent of the Minister of Health. The laboratory advised the supplier not to let any more Radon Sparklets

Bulbs — a product which its English manufacturer had stopped making — leave the premises, pointing out that it was not the radon produced by the bulbs that was likely to cause harm, but that 'in making the radon-activated water the contents of the bulb may become dislodged and enter the water used, making possible the ingestion of radium'.[27] The amount of radium contained in the bulbs was considered enough, over time, to cause death if lodged permanently in the human body. Apart from the danger of radium being ingested by the user, the spent bulbs were also considered a serious health hazard because they contained unsealed radium that could become widely distributed.

Natural environmental radiation, or background radiation, was also the subject of increasing research. Ernest Marsden continued his own investigations into environmental radioactivity, with much of his interesting and unusual research attracting coverage in the daily press. His most publicised findings came from his research into Niue Island, where a DSIR Soil Bureau study had shown that the island's soil had unusually high levels of radioactivity. Marsden, inspired to investigate further, found the radioactivity of food grown on the island to be up to 100 times normal. His findings caused quite a stir internationally, with the popular press picking up on Marsden's assertions that Niueans were a master race. Not only were they taller, much happier and less prone to disease than other races, he said, but selective breeding had led to the population building up a resistance to radiation which would be advantageous in the event of a nuclear war. Despite criticism of his theory, Marsden persisted, stating in 1962: 'My contention that the people of Niue Island would be better off in a nuclear war than the rest of us is a good story and I'm sticking to it!'[28]

Another of Marsden's high-profile projects was his investigation into the radioactivity of tobacco. By the 1960s, links between cigarette smoking and lung cancer were evident. Marsden saw the striking increase in British deaths from lung cancer as being possibly linked to increased imports of Southern Rhodesian tobacco, which he had found to have high levels of polonium-related radioactivity.* In 1965, at Marsden's request, the DSIR's chemistry division developed a new type of cigarette filter designed to reduce the amount of polonium inhaled when smoking cigarettes.

---

* Polonium is a rare but naturally occurring radioactive element, of atomic number 84, first discovered by Marie Curie in 1898.

Marsden, with his daughter and wife, at the University of Manchester for the 1961 Rutherford Jubilee International Conference. Marsden was held in high regard by his British peers, and was elected president of this gathering of 500 of the world's leading physicists. Lady Marsden Collection, PAColl-0091-1-013, Alexander Turnbull Library, Wellington, New Zealand.

Despite his seemingly eccentric scientific pursuits, Marsden maintained his international scientific connections and was held in high regard by the physics community. While working on his retirement projects he corresponded with some of the top Commonwealth nuclear scientists — including John Cockcroft and William Penney in the United Kingdom, and Charles Watson-Munro in Australia — using his connections to call in favours for advice or equipment that might otherwise have been difficult to obtain. (In return, Marsden was known to send eminent scientists, including Cockcroft, parcels of New Zealand lamb, to arrive just in time for Christmas.) In 1961 he was invited to be President of the Rutherford Jubilee International Conference in Manchester, a gathering of 500 of the world's leading physicists to commemorate the fiftieth anniversary of the discovery of the atomic nucleus.

RADIOACTIVE WASTE

The oceans were once considered a suitable dumping site for industrial waste, including radioactive waste. In 1955, when it was expected that New

Zealand would soon have nuclear power stations for electricity generation, some scientists suggested that any atomic waste generated could be sealed in concrete blocks and dropped into the Kermadec Trench. At this time, oceanic trenches were believed to be troughs of stagnant water and were therefore suitable burial grounds. Even so, this was considered to be only a short-term solution, until the technology existed to shoot the nuclear waste into space 'and let it revolve around the world, doing no harm to anyone or anything'.[29] Suggestions like this did not arouse concerns, as consigning radioactive waste to the deep ocean trenches or outer space meant it was considered to be far from having any harmful influence on humanity.

At the Second United Nations Conference on the Peaceful Uses of the Atom in 1958, Soviet scientists warned that this practice could be dangerous. They dealt particularly with the Kermadec Trench, and from their observations concluded that the water in the trench was in contact with the general oceanic circulation and was not, therefore, a safe place to dump radioactive wastes. By this time, it was estimated that the United States had dumped about 10,000 curies* of nuclear waste into the Atlantic Ocean as well as some into the Pacific Ocean. The United Kingdom admitted to dumping 500–600 curies of nuclear waste in more than 2000 fathoms (about 4000 metres) of water in the Atlantic, an amount of radioactivity that was described at the 1958 conference as considered 'perfectly trivial'.[30] Low-level waste was also being pumped into the Irish Sea as effluent from the United Kingdom's Windscale nuclear power plant.

While the marine disposal of radioactive waste was an issue for discussion by the 1950s, it was still acceptable. In 1956, the United Nations held its first conference on the Law of the Sea, out of which came four conventions covering territorial zones, the continental shelf and the high seas. Article 25 of the 1958 Convention on the High Seas, which finally came into force in 1962, stipulated that 'Every State shall take measures to prevent pollution of the seas from the dumping of radio-active waste.' The convention did not outlaw the dumping of radioactive waste in the sea; it was just concerned with pollution as a result of such dumping, and, as marine disposal was still considered the best thing to do with radioactive waste, New Zealand continued to dump radioactive waste into the sea until 1976.

---

* A curie is a measure of radioactivity, roughly equivalent to the activity of 1 gram of radium-226. Today, radioactivity is more commonly measured in the SI unit becquerels.

The radioactive isotopes New Zealand imported were used in medicine and industry, and created only low-level radioactive waste. While the activity of the waste was insignificant in comparison to the level of waste generated in nuclear reactors, it still had to be disposed of safely. This low-level waste took the form of contaminated articles, such as gloves and equipment, and used or damaged pellets of isotopes like caesium-137, cobalt-60, strontium-90 and radium. For many years, this waste was dumped into the Hikurangi Trench, just east of Cook Strait. Sealed radioactive sources were embedded in concrete in 20-litre steel drums. The 1972 Convention on the Prevention of Marine Pollution by Dumping of Wastes and Other Matter was a global convention to protect the marine environment from human activities. With the adoption of the requirements of the convention in New Zealand's Marine Pollution Act 1974, a special dumping permit was required for such disposals. This procedure was followed once in 1976, after which New Zealand's waste was stored at the National Radiation Laboratory in Christchurch.

NUCLEAR WASTE (SEALED SOURCES) DUMPED AT SEA UP TO 1976

NUCLIDE	TOTAL DISPOSALS AT SEA UP TO 1976
Cobalt-60	23 GBq
Strontium-90	3.2 GBq
Caesium-137	1.0 TBq
Radium-226	17 GBq
Americinium-241	190 GBq

Note: Radioactive waste disposed at sea in New Zealand, with activity measured in becquerels (Bq), the SI unit of radioactivity that replaced the curie (Ci). One becquerel is equivalent to one nucleus decay per second. (GBq = gigabecquerel; TBq = terabecquerel.)

Source: M. K. Robertson, *Radioactive Waste Disposal — Policies and Practices in New Zealand*, National Radiation Laboratory Report 1996/2, National Radiation Laboratory, Christchurch, Feb. 1996.

New Zealand was also briefly involved in the transit of American radioactive waste through New Zealand. When the United States installed a nuclear power station at McMurdo Station in New Zealand Antarctic Territory in 1961, the Antarctic Treaty required that radioactive waste be returned to the United States for disposal. New Zealand had earlier been involved in the initial transport of nuclear material to the Antarctic reactor in the summer of 1961–62. The reactor (known locally as 'Nukey-Poo') was installed at the American base on Ross Island, where, previously, half of all freight shipped to the base was fuel for generators, space heaters and vehicles. The reactor's radioactive fuel cores and a start-up source had to be replaced

McMurdo Station's nuclear reactor, known locally as 'Nukey-Poo', sat at the base of Observation Hill and powered the American Antarctic base from 1962 to 1972. It leaked, and after it was decommissioned thousands of tonnes of contaminated soil had to be shipped out of Antarctica and back to the United States. Courtesy Pam Landy.

several times over the life of the reactor, so uranium-235, as well as neutron sources of mixed polonium and beryllium, were shipped through Lyttelton on their way to McMurdo Station. Ships carrying the spent fuel, along with many barrels of low-level radioactive waste, also stopped in Lyttelton on their way back to the United States.

Under the Radioactive Substances Act 1949 the Department of Health had the right to enter any ship carrying radioactive substances, but in a 1958 agreement between the two countries New Zealand had waived all rights to inspect any United States property in transit through New Zealand on its way to Antarctica. The Dominion X-Ray and Radium Laboratory expressed concern over the safety of the reactor fuel and neutron source being transported through New Zealand ports, and the safety of the return of spent fuel elements and of waste radioactive material. Nonetheless, they agreed to waive inspection rights provided they were given a schedule of any nuclear materials and waste being shipped through New Zealand, and assurance that the shipments were in accordance with the IAEA Regulations for the Safe Transport of Radioactive Material.

The Antarctic reactor was plagued with problems, however, and by 1967 earlier plans to add nuclear power stations at American bases at Byrd Station and South Pole Station had been shelved. After the reactor had been operating for a decade, a routine inspection in 1972 revealed a crack in a water tank used to provide radiation shielding. Further investigations revealed that some of this water had leaked into surrounding insulation and the soil below the reactor. As it was too expensive to import the equipment required to more fully test and repair the damage, the reactor was decommissioned. To comply with Antarctic Treaty provisions, any soil showing traces of radioactive contamination was removed along with the dismantled reactor. In the years following the reactor shut-down, thousands of tonnes of contaminated soil from underneath and around the site of the reactor were removed and shipped back to the United States, through New Zealand. The main radioactive contaminant was caesium-137, but the level of contamination was mostly so low that the shipments were exempt from approval or inspection by the National Radiation Laboratory.

This didn't stop public concern. In 1975, Robert Mann, an Auckland biochemistry lecturer and director of the Environmental Defence Society, brought the shipments of radioactive soil to the attention of the media, who responded with headlines like 'Hush-Up Over Deadly Cargo: Minister Asked to Explain'. One of Mann's concerns was the New Zealand Electricity Department's plans for a nuclear power station to provide electricity to Auckland. Mann told *The Christchurch Star* that the Government's 'casual' attitude to the nuclear reactor and its 'potential dangers' on Antarctic soil did not bode well for 'attitudes and policies regarding its plans to set up a nuclear plant here on the mainland in 1979'.[31] The Minister of Health of the Labour Government, Thomas McGuigan, who was also the MP for Lyttelton, downplayed the significance of the shipments, saying there was never any risk and accusing Mann of using 'scaremongering tactics'.[32]

Before the waste was stored at the National Radiation Laboratory, some of New Zealand's own nuclear waste was disposed of in landfills, like contaminated laboratory equipment from Victoria University buried at Wilton Tip (now Ian Galloway Park) in the early 1960s. As Rachel Barrowman recounts in her history of Victoria University of Wellington,[33] in 1961 a technician trying out a new Geiger counter discovered certain rooms in the university's physics department, which had been used by early physicists such as Charles Watson-Munro, recorded high levels of radioactivity, possibly as a result of careless handling of radium-226. Technicians from the Dominion X-Ray and Radium Laboratory surveyed the department and removed some contaminated material, which was encased in concrete and dumped at sea, as was standard practice at the time. Two years later a new radiation monitor revealed further evidence of contamination, and a more comprehensive clean-up was ordered. Radioactive contamination was discovered on furniture, equipment, floors, walls, dust and papers. Meanwhile, in 1963 Ron Humphrey, a physics lecturer who had been working in the contaminated rooms, died of leukaemia after a two-year illness, and his family were suing the university for compensation. The department was carefully decontaminated, and all contaminated material, including a large workbench, was removed to a nearby garage in Waiteata Road, pending a possible court case over Humphrey's death. When the case was settled out of court, the material was trucked to Wilton Tip. The cause of Humphrey's leukaemia was not determined, but post-mortem analysis of his bones — the place where ingested radium becomes concentrated — showed his radium levels were within the normal range.

The story made front-page news in Wellington's *The Dominion* when the story broke in 1963, with a headline 'Radiation Danger Discovery'.[34] The reporting, however, was straightforward: the Dominion X-Ray and Radium Laboratory was treated as an authority on the matter and there was no speculation that they might be withholding information. Waiteata Road residents were concerned, though, and complained to the Department of Health. Turbott assured them that even if a person were in the same room as the materials in the garage, they 'would not be hazardous to a person not in direct contact with them'.[35] Attitudes were to change over the coming decades, however. When the same issue attracted media attention during building renovations in 1988, by which time New Zealanders had embraced the nuclear-free policy, the story again resulted in front-page headlines — this time they read 'Radiation leak: official cover-up' — as well as reports and interviews on radio, television and in newspapers around the country.[36]

NUCLEAR CIVIL DEFENCE

In the United States, preparation for nuclear attack led many cities to build fallout shelters beneath city streets and inspired homeowners to build backyard nuclear bunkers. New Zealand was never considered to be a primary nuclear target, but safety from nuclear explosion or fallout was an issue for New Zealand's military and civil defence. From a twenty-first century perspective it is clear that the greatest threats to twentieth-century New Zealand were earthquakes, cyclones and floods. But it was preparedness

─ ATOMIC BOMBING OF WELLINGTON CITY. ─

KHANDALLAH

NGAIO

WADESTOWN

THORNDON

NORTHLAND

| ZONE 1 VERY HEAVY | ZONE 2 HEAVY | ZONE 3 MODERATE |

LAMBTON HARBOUR

KARORI

KELBURN

CITY CENTRAL

TE ARO

ORIENTAL BAY

EVANS BAY

MITCHELLTOWN

HATAITAI

BROOKLYN

MIRAMAR

NEW TOWN

KILBIRNIE

BERHAMPORE

LYALL BAY

ISLAND BAY

THE 4 ZONES INDICATE THE EFFECT OF THE ATOMIC BOMB ON WELLINGTON, WHEN EXPLODED 880 YARDS

ABOVE THE RAILWAY STATION

SCALE 1 MILE = 4 INCHES

for a nuclear attack that prompted New Zealand's first civil defence schemes and the eventual establishment of a Ministry of Civil Defence.

The New Zealand Government's White Paper *Review of Defence 1958* stated that:

> . . . The safeguarding and educating of the civil population against the nuclear effects of war must, for the first time, become an essential part of national defence plans. The geographical position of New Zealand no longer affords the country security from the worst impact of a global conflict. A nuclear war and the hazards to civilian population of radioactivity will not necessarily be confined to the countries of the main combatants. Radioactivity knows neither frontiers nor distance and the contamination of nuclear weapons could assume world-wide proportions.
>
> . . . The defence plan must also take into account the possibility of a direct attack on this country with nuclear or non-nuclear weapons. Even a single submarine with guided missiles would offer a considerable threat to our shores.[37]

By now, even countries like New Zealand, remote from the northern hemisphere nuclear powers, were feeling threatened: intercontinental ballistic missiles had first been launched in 1957, the Soviet Union had launched Sputnik 1, the first artificial satellite, and the United States was building submarines capable of firing nuclear missiles. As we saw in Chapter Three, Nevil Shute's 1957 novel *On the Beach* had had a profound impact on society and had fuelled the growing awareness that the southern hemisphere, New Zealand included, would not escape the effects of a nuclear conflict.

When the Ministry of Civil Defence was set up in 1959, its primary concern was the threat of nuclear attack. A 1959 publication, *Civil Defence in New Zealand*, declared the most likely nuclear targets to be Auckland, Wellington, Christchurch and Dunedin. The threat from nuclear weapons was considered to be very real, and the booklet described the effects of nuclear weapons as devastating 'property and personnel by heat, blast and direct radiation over a circular area' with 'residual radiation from the fall-out . . . fatal to persons exposed to it without protection for 24 to 48 hours'. After that, it said, 'exposed persons would be made ill from radiation sickness

Opposite: This Army Department map (dating from about the late 1950s) shows fallout zones for a nuclear bomb-drop on Wellington city. AD 66 34 Misc 26, Atomic bombing of Wellington city, Archives New Zealand, Wellington, New Zealand.

and some might die'. While evacuation of potential targets was considered the civil defence priority in a nuclear attack, the booklet also mentioned the need for fallout shelters 'of relatively simple type such as slit trenches and Anderson-type frames [a type of Second World War bomb shelter] partially sunk in the ground, with a cover of 3ft of earth'.[38] Shelter policy, it advised, would be the subject of future planning and advice. After 1963, though, when the Limited Test Ban Treaty was signed, the ministry's functions began to tend towards natural-disaster preparedness rather than nuclear attack.

Nuclear ship visits were also a civil defence issue. In 1958, Roth said that the hazards arising from possible collisions or other accidents meant that 'only the most compelling economic or military reasons would justify the entry of a nuclear powered ship into the Port of Auckland'.[39] Even so, Walter Nash's Labour Government in 1960 accepted a visit of the USS *Halibut*, an American nuclear submarine, to Auckland and Wellington. This first nuclear ship visit was followed in 1964, under Keith Holyoake's National Government, by visits from American nuclear-powered cruisers USS *Long Beach* and USS *Bainbridge*, and the nuclear-powered aircraft carrier USS *Enterprise*.

In 1967, the National Radiation Laboratory's Jim McCahon wrote to the Department of External Affairs on the issue of nuclear ship visits, pointing out that nuclear-propelled vessels 'carried a very small risk of an accident which could affect people and installations on the shore' and suggested it was reasonable to require indemnity for the entry of a nuclear-powered ship into New Zealand waters.[40] Holyoake's Government advised the United States in 1971 that a condition of future visits from nuclear-powered ships would be that the United States agree to accept liability in the event of a nuclear accident, and a New Zealand Code of Practice for Nuclear Powered Shipping was prepared by the New Zealand Atomic Energy Committee. But by the time the United States agreed to accept liability, in 1974, a new Labour Government, under Bill Rowling, chose to continue the ban on nuclear-powered ships.

New concerns about New Zealand becoming a nuclear target arose in 1968, when *The Press* revealed plans that United States Navy engineers, working with officials from New Zealand's Lands and Survey Department, the Ministry of Works and the Post Office, had inspected three possible sites for a radio transmitter for the Omega navigation system: in the Lake Sumner and Lake Pearson areas, and in the Omarama district of North Otago.

The purpose of the new system was to provide American aircraft, ships or submarines equipped with an Omega radio receiver with data on their exact geographical position anywhere in the world, to within about one or two nautical miles; in the days before global positioning satellites, this was a great technological advance. *The Press* reported that the Omega system would operate on VLF (very low frequency), meaning the signals could travel very long distances, with the system working with only eight stations worldwide. The University of Canterbury student newspaper, *Canta*, responded with three full pages on the story, claiming that 'the New Zealand Government plans to place us among one of the eight most important [nuclear] targets in the world'. 'Until now,' said *Canta*:

> . . . New Zealanders have drawn some comfort from the knowledge that in
> the event of nuclear war, there is no apparent reason why any hostile nation
> might select so small, relatively sparsely populated, and unimportant a land
> for destruction. At one stroke, the New Zealand Government plans to place
> us among one of the eight most important targets in the world. . . . As a hostile
> nation would be unable to locate U.S. polaris submarines, the only effective
> way to reduce American firepower would be to eradicate as rapidly as possible
> the known 8 stations by which they navigate. . . . From the moment such a
> station reached completion, Intercontinental missiles with atomic warheads
> will be aimed to home on New Zealand, since this would be the only way to
> remove the threat from the world roving polaris fleet. . . . If it is the will of New
> Zealanders to be the target during the first few hours of atomic attack, then
> there will be no opposition to this plan. If enough of you believe that it should
> be opposed, then let your strongest voice be heard over this land, before this
> thing is done.[41]

Canta also reported that Christchurch airport would be a target, and 'the sparsely populated hinterland of Canterbury would be reduced to an arid desert. Those not killed by the initial blast heat or radiation would die from fallout spreading from the explosions in the alps and at Christchurch Airport. Christchurch itself would be completely destroyed.'[42]

People responded to *Canta*'s call for action and Omega became a national issue. Suddenly the idea that New Zealand might become a nuclear target seemed very real. Students demonstrated against the proposed station, people marched in protest, and the Labour opposition leader, Norman Kirk, spoke out against Omega. Meanwhile, debate raged about whether the system was intended for military or commercial shipping. Either the protestors' voices were heard or New Zealand's terrain proved too challenging — the proposed station was not built in New Zealand; it was eventually built in Victoria, Australia.

THE EVOLVING AWARENESS OF RADIATION RISKS

In the 1950s, New Zealanders became aware of the health risks posed by radioactive fallout from bomb testing, and the public began to be wary of the dangers posed, for example, by strontium contamination of milk. But this fear of radiation dangers from fallout was associated with the testing of nuclear bombs and did not usually extend to a fear or suspicion of nuclear power or nuclear science and medicine. Despite growing evidence of the dangers of all exposure to radiation, workers were happy to line up for a free x-ray to screen for tuberculosis, parents let their children's feet be irradiated by pedascopes in shoe shops, radon-irradiated water continued to be offered for sale, and watchmakers hid their supplies of radioactive paint from authorities. By the end of the 1960s, however, public awareness had grown and demand for these potentially dangerous novelties waned at the same time as government imposed restrictions on their use. In hospitals, the non-prescriptive legislation was working: average recorded radiation exposure by medical staff, as recorded on film badges, declined steadily from 1954 to 1967.[43]

Concerns about fallout led to public action, which prompted the Government to be more open about fallout information. The National Radiation Laboratory — the public's trusted source of advice and information on radiation issues — began publishing quarterly fallout statistics in the 1960s. But in other applications of nuclear science and medicine, it was not public demand but government officials responding to the latest scientific evidence or to international law that led to the implementation of more stringent regulations. The Department of Health took the lead in advising the medical profession and the public on safety measures, and had to take measures to control over-enthusiastic users of radioactive materials who chose to continue using them — radium-activated paint and radon-infused water — against medical advice.

6

★

Atoms for Peace
Nuclear science in New Zealand in the atomic age

A country backward in nuclear science can only stumble blindly in the atomic age, ignorant of opportunities, deficient in technique and the pawn of countries more advanced. — J. WILLIAMS, REPORT ON DEVELOPMENT OF NUCLEAR SCIENCES IN NEW ZEALAND, 19 JULY 1956[1]

A new source of power to light the homes of the people and turn the wheels of industry; an order to build a ship that will cross the seas without coal or oil fuel. This atomic age is indeed beginning to show signs of an assured future. — THE DOMINION, 17 OCTOBER 1956[2]

In the 1950s, the United States launched its Atoms for Peace programme — an international programme initiated by President Dwight D. Eisenhower that promoted so-called peaceful uses of atomic technology. New Zealanders bought into Eisenhower's utopian vision of an atomic future, and when the United States nuclear submarine USS *Halibut* visited New Zealand in 1960, thousands of Aucklanders and Wellingtonians flocked to the ports to welcome the vessel — the world's first nuclear-powered submarine. New Zealand's National Film Unit included the visit in their regular *Pictorial Parade*, describing the 'sleek dark shape' entering Wellington Harbour in the

The world's first nuclear-powered submarine, the USS *Halibut*, visited Auckland and Wellington in April 1960. The nuclear submarine was greeted as a positive symbol of a future in which atomic energy would be used for peaceful purposes. Courtesy APN New Zealand.

early morning and marvelling that 'somewhere inside her long grey hull a small atomic reactor provides enough power to take her round and round the world'.[3] *The New Zealand Herald* described the 'silent, slate-grey' submarine that arrived at Devonport as an 'impressive sight', marvelling that she had travelled most of the 3900 miles (6200 kilometres) from Pearl Harbor under water. 'Ferry passengers gaped and New Zealand sailors slipped away from their jobs to peer at her,' the paper continued.[4]

Although opposition to Pacific nuclear bomb tests was growing, New Zealanders remained enthusiastic about what the United States promoted as 'peaceful' uses of nuclear technology and welcomed the USS *Halibut* as a positive symbol of an atomic future.

NUCLEAR SCIENCE IN NEW ZEALAND

After Ernest Marsden left the DSIR, the organisation was run by agriculturalists — Frank Callaghan from 1947 to 1953, then Bill Hamilton

156

until 1971. With Marsden gone, there was no one to champion nuclear projects; what's more, the organisation was operating under a budget slashed by Sidney Holland's National Government and any new proposals were subject to intense economic scrutiny. By the mid-1950s, many of the young DSIR scientists who had worked on the wartime Manhattan Project and on the nuclear reactor projects in Canada and the United Kingdom had, because of limited opportunities in New Zealand, moved overseas or into other areas of research.

Charles Watson-Munro, who led the New Zealand teams working on the Canadian and British nuclear reactor projects (ZEEP and GLEEP), left the DSIR in 1951 to be professor of physics at Wellington's Victoria University College. He did not stay long: in 1955 he moved to Sydney to become chief scientist with the Australian Atomic Energy Commission. Australia's atomic energy projects were now well in advance of New Zealand's: a uranium mining industry was exporting to the UKAEA, and a £5.5 million atomic energy programme had included the construction of a heavy-water moderated, enriched-uranium research reactor (HIFAR) at Lucas Heights near Sydney. George Page and Gordon Fergusson, both part of Watson-Munro's team that had worked on GLEEP, continued working on nuclear sciences in the DSIR's Dominion Physical Laboratory, although Page soon left New Zealand to join Watson-Munro and fellow New Zealander Cliff Dalton at Lucas Heights.

After Marsden left his position as head of the DSIR, another New Zealand scientist took a leading role in developing the field of nuclear science, taking it in a new direction. By becoming a world expert in the nascent field of radiocarbon dating, Athol Rafter established an international reputation for the DSIR's nuclear sciences team, and set the direction for the future research of what became the Institute of Nuclear Sciences. In 1948, Rafter, a DSIR chemist, had been sent with Gordon Fergusson on a trip to the United States and the United Kingdom to learn about radiochemistry. Back in New Zealand, Rafter later recalled, 'no one really knew what to do with us. Nuclear science was a very young baby clothed in mysticism and nuclear annihilation. Any expenditure on nuclear science was of necessity a major expenditure, and any work involved hazards that the people of New Zealand little understood.' Rafter and his colleagues, however, managed to assemble enough equipment to provide nuclear science services to medicine, agriculture and industry. Working in a shed 'that looked like an outhouse' Rafter prepared radioactive isotopes for use around New Zealand.[5] Page, working in the attic of the Dominion Physical Laboratory, built a mass spectrometer, and Fergusson assembled equipment to measure natural background radioactivity.

The nuclear sciences team applied their expertise to investigating nitrogen metabolism in apple trees; finding the origin of sulphur in New Zealand coals; measuring radioactivity levels in waters, gases and soil samples; and surveying the heavy-water content of New Zealand waters. Rafter also began experimenting with the new technique of radiocarbon dating, which calculates the age of organic material by comparing the ratio of radiocarbon atoms (carbon-14) to regular carbon atoms (mostly carbon-12) in a sample with known radiocarbon:carbon ratios from different time periods. Rafter was prompted to investigate the new technique by Callaghan, who requested that Rafter attempt to date the age of New Zealand's volcanic ash showers to 'stop the geologists arguing'.[6] Rafter used radiocarbon dating first to date pieces of organic matter from ash deposits in the North Island and to date moa bones. After difficulties with the process used by Willard Libby, the American pioneer of radiocarbon dating, Rafter worked with Fergusson and a DSIR team to perfect a more reliable method of radiocarbon dating using carbon dioxide gas rather than solid carbon. The new technique was very successful: the carbon dioxide method soon became the standard procedure, and Rafter and his team established an international reputation for radiocarbon dating.

An *Evening Post* article in 1952 described New Zealand's nuclear scientists as using radioactive isotopes from Harwell, and working 'quietly, often in inadequate laboratories and with makeshift equipment which astonishes visiting scientists' where they were 'doing very valuable work in the fields of industry, agriculture, medicine and historical research'.[7] Outside of the DSIR, other New Zealand research laboratories were making use of radioactive and stable isotope tracers: the Department of Agriculture's Animal Research Laboratory to determine the effect of cobalt on animal metabolism; the New Zealand Fertiliser Manufacturers' Research Association to study fertiliser uptake in plants; and the Otago University Medical School to study thyroid function. Radioactive isotopes were also being used in industry, for example to trace welding faults at the Maraetai power station on the Waikato River.

In the universities, Auckland University College had constructed and was using a low-energy linear accelerator — a machine that can accelerate subatomic particles and ions to high speeds — for research, and the University of Otago was constructing a high-energy Van de Graaff accelerator. Victoria University College and Auckland University College were studying cosmic rays and radioactive contamination of the atmosphere, and there was work on low-energy x-rays at Canterbury University College.

Nuclear science techniques practised by New Zealand scientists were now established in many areas of New Zealand industry and agriculture. But the most extensive use of radioactive isotopes was still in medicine. In radiation therapy an isotope of cobalt, cobalt-60, was now used in preference to radium as it was a cheaper source of radioactivity for treatment. Other radioactive isotopes for diagnosis and therapy were prepared for hospital use by DSIR scientists from bulk shipments received from overseas. With no local source of the radioisotopes needed for research and industrial, agricultural and medical applications — from the cobalt-60 and radium-228 used in cancer treatment to the caesium-137 and phosphorus-32 used in industry and agriculture — New Zealand had to import them, initially from the United Kingdom, the United States and Canada, and, from the 1960s, from the HIFAR nuclear reactor in Australia. Because of the limited useful life of most isotopes, they had to be brought to New Zealand by air. But some particularly short-lived isotopes, such as fluorine-18, which was used overseas in dental examinations and had a half-life of just 109 minutes, could not be used in New Zealand as most of the isotope would decay in the time taken to transport them from Australia.

EISENHOWER'S ATOMS FOR PEACE SPEECH

The United States began promoting peaceful uses of nuclear technology in the 1950s. In December 1953, the United States President, Dwight D. Eisenhower, addressed the United Nations General Assembly with a speech promoting the establishment of an international atomic energy agency that could stockpile fissionable materials for use by non-nuclear powers for peaceful purposes. He called for atomic energy to be applied 'to the needs of agriculture, medicine and other peaceful activities' and specifically 'to provide abundant electrical energy in the power-starved areas of the world'.[8]

While the nuclear powers discussed the establishment of an international atomic energy agency, the United States started to fulfil the promises made in Eisenhower's Atoms for Peace speech by offering assistance to other

In 1953 American President Dwight D. Eisenhower addressed the United Nations General Assembly with a speech calling for atomic energy to be applied 'to the needs of agriculture, medicine and other peaceful activities'. United Nations.

countries to construct small-scale nuclear reactors, sponsoring international scientific conferences on nuclear science, and providing technical information and training programmes. While planning proceeded for a United Nations conference on peaceful uses of the atom, the United States began to sign bilateral agreements to provide non-nuclear countries with information on the design, construction and operation of research reactors. The United States promoted its Atoms for Peace programme as being motivated by a desire for world peace and prosperity, but the country's National Security Council was clear that any bilateral agreements the United States entered into regarding provision of atomic energy or nuclear technology should seek to promote the United States' own atomic energy interests, and any nuclear materials provided by them to another country must be returned for reprocessing in the United States.

By promoting bilateral agreements ahead of the multilateralism soon to be introduced in an international atomic energy agency, the United States was able to maximise its own influence and control over the Western world's nuclear industries, and reap any benefits from research advances made in its partner countries. The main focus of the United States' nuclear research at this time was, of course, nuclear weapons, and at the same time as promoting the 'peaceful uses' of nuclear technology the United States was continuing to test larger and more devastating nuclear weapons in its Pacific testing zones. Less than three months after Eisenhower's Atoms for Peace speech, the United States exploded a massive hydrogen bomb — still the fifth largest nuclear explosion in history — at Bikini Atoll. The unexpectedly large explosion caused widespread radioactive contamination and the death of at least one person.

The biggest advances in 'peaceful' nuclear technology were now taking place in the United States where Willard Libby had developed the technique of radiocarbon dating and Robert Van de Graaff had developed a powerful new type of particle accelerator. New Zealand was physically and scientifically remote from these advances, but still liked to acknowledge the role of Ernest Rutherford in the scientific co-operation that had led to the birth of nuclear physics. While the United Nations was working on the United States' Atoms for Peace conference proposal, New Zealand had a representative on the United Nations Security Council, Leslie Munro. In a 1954 speech to the First Committee of the General Assembly, Munro drew attention to New Zealand's own Ernest Rutherford and his role in the development of knowledge that would later lead to the peaceful uses of atomic energy, and to the international co-operation in atomic research that had persisted up until the Second World War. 'Dare I express the hope', Munro said:

> . . . that it may now be possible, slowly but surely, to revert to the basis of international cooperation on which our atomic science was constructed. Is it too much to hope that before long the Rutherfords and Kapitzas of present-day atomic physics may again join in seeking solutions to the many mysteries which still remain? No matter how great its resources, no matter how advanced its technology, no matter how numerous or well trained its scientists, no one nation could hope by itself to match the pooled efforts of a group of nations in this field. On the other hand, scientists from even the smallest

countries, which may have little to offer by way of raw materials or industrial capacity, can make vital contributions.[9]

As well as calling for the resumption of international co-operation among atomic scientists, Munro expressed New Zealand's support for the establishment of an international atomic energy agency and a proposed scientific conference on the peaceful uses of the atom.

The United States was active in seeking a bilateral agreement with New Zealand. In 1954, New Zealand was one of several countries visited by an American team offering government-level co-operation with atomic energy projects. The Department of External Affairs was enthusiastic at the prospect of cementing New Zealand–American relations through signing a bilateral agreement under the Atoms for Peace programme, but how did the government agencies that would be most involved in a nuclear reactor for research or power generation — the DSIR and the State Hydro-electric Department — react? In contrast to the External Affairs officials, who were focused on strengthening New Zealand's relationship with the United States, representatives of the government's scientific and engineering agencies were mindful of the hidden costs of the United States' offers. Arthur Davenport, Secretary of the State Hydro-electric Department, in May 1955 told the Secretary for External Affairs, Alister McIntosh, that 'the construction of a research reactor in New Zealand would not assist in the practical application of nuclear energy in the generation of nuclear power. If the construction of such a research reactor is an essential part of the bilateral agreement with USA, this Department is not prepared to recommend the signing of the agreement.'[10] Hamilton was also concerned about how the proposed bilateral agreement might affect scientific relations with the United Kingdom. He suggested to McIntosh that any exchange of information between the United States and New Zealand be restricted to unclassified data as 'the receiving of even a small amount of classified information might limit freedom of publication of original work done in New Zealand, or work based on information obtained in an unclassified form from other sources, such as the United Kingdom'.[11]

The United States' standard bilateral agreement became more worthy of New Zealand's consideration in 1955 when it was revised to include funding for half the price of a nuclear reactor (with a ceiling of £125,000, or US$350,000). The DSIR, however, was still coping with a reduced budget, and advised the Department of External Affairs that New Zealand was not yet ready for such an agreement, noting that the 'high capital cost involved and the cost of maintenance would be out of all proportion to the benefits which would accrue, particularly as we are able to secure for our present limited needs all the radioactive isotopes which New Zealand requires for research purposes in agriculture, medicine, biology'.[12] While the United States now had the scientific lead in nuclear technology, there were still many personal and professional links between New Zealand and British scientists; the DSIR said that were New Zealand in the future to make a decision to construct a research reactor, it would be preferable to do so through existing links with the United Kingdom.

As well as wanting to maintain existing links with British science and scientists, there seems to have been a degree of mistrust of the United States amongst New Zealand's science administrators. The United Kingdom was concerned about the possibility of New Zealand buying an American reactor, saying that they did not want New Zealand to buy an inferior or outdated reactor (they believed their technology was superior) and offering their help. The DSIR was happy, however, to accept a technical library from the United States Atomic Energy Commission. When the American ambassador presented the library to Holland in October 1955, the New Zealand prime minister made a connection between Ernest Rutherford and the latest nuclear technology, proudly noting that 'in a sense all the scientific advances surveyed in this assembly of material had their beginnings in this country of ours'.[13] Rutherford, of course, did no work on nuclear science in New Zealand but, once again, the New Zealand Government was happy to claim domestic links with this exciting new field of science.

ATOMS FOR PEACE CONFERENCE 1955

The first United Nations International Conference on the Peaceful Uses of Atomic Energy — commonly known as the Atoms for Peace conference — was held in Geneva from 8 to 20 August 1955, ten years after the bombing of Hiroshima and Nagasaki. The conference was the first large-scale meeting of scientists from the West and from Communist-bloc nations since the Second World War, and was marked by the release of much previously secret or highly classified information. The main scientific concerns of the conference

— which presented nuclear energy as the only long-term solution to society's energy needs — were the generation of electricity through nuclear reactors, and the use of radioisotopes in medicine, biology, agriculture and industry.

At the conference, the United States re-stated its offers of half-price nuclear reactors and supplies of fissile materials to countries undertaking research in nuclear physics and adding to the sum of world knowledge about peaceful uses of the atom; and offered opportunities for advanced training for scientific and engineering graduates from countries planning to construct atomic power stations.

Although the United States' offers of assistance were outwardly generous and very well publicised, it was not the only country offering assistance in the field of nuclear technology. The Soviet Union gave technical assistance to China, Poland, Czechoslovakia, Romania and East Germany to help them set up nuclear physics research facilities. The United Kingdom, whose atomic energy programme was initially more advanced than that of the United States, and who was unencumbered by any secrecy clauses such as in the United States 1946 Atomic Energy Act, had already been offering informal technical assistance to other European countries and hosting visitors from New Zealand and other Commonwealth countries. This Commonwealth scientific co-operation dated back to the Second World War, when a system of formal scientific liaison officers was set up. One of the aims of the network was to share information and pool research results — particularly in the area of defence science — a scheme that the United States found threatening and had opposed. In June 1946, after a Prime Ministers' Conference held in London, an informal Commonwealth Conference on Defence Science was held to distribute research and development in defence science. The Dominions, it was decided, could assist Britain by providing scientists and engineers, who were in short supply in the United Kingdom, in return for post-graduate training; the arrangement in which New Zealand scientists trained at the Atomic Energy Research Establishment was of as much benefit to the United Kingdom as it was to New Zealand.

Hamilton led the New Zealand delegation to the 1955 Atoms for Peace conference. Along with government representatives from the State Hydro-electric Department, the Ministry of Works and the Department of External Affairs were Francis Farley, an Auckland University College physicist who had research experience at ZEEP in Canada and at Harwell in the United Kingdom, and Darcy Walker, physics professor at Victoria University College.

The delegates reported on the conference on their return. While they all accepted the need for New Zealand to have nuclear power in the future, and agreed that people would have to be trained in preparation, they disagreed over whether or not New Zealand should establish a nuclear energy research programme.

Again, it is interesting to note the difference in stance between External Affairs officials, who were clearly seeking to strengthen relations with the United States — by now the Western superpower and one of New Zealand's partners in the ANZUS Treaty — and the more insular responses from New Zealand's scientific and engineering communities, who focused on what they saw as New Zealand's immediate needs. In his report on the conference, Lloyd White, from New Zealand's High Commission in London, noted that 'a country's political weight will henceforth be judged, at least in part, by its participation in the atomic field. If we want to maintain an influential position among nations . . . we must take part in this progress.' He went on to argue in favour of an atomic research programme in New Zealand, noting that while it was too early to place an order for a nuclear reactor for atomic energy production, it was not too early 'to be thinking about training atomic technicians and accumulating a body of knowledge on the sort of atomic reactors which would be best suited to New Zealand conditions'.[14]

In his book on New Zealand foreign policy during the nuclear age, Malcolm Templeton described White's boss, External Affairs head Alister McIntosh, as 'ever-cautious' and 'no great lover of scientists',[15] but McIntosh did recognise the potential value to New Zealand of co-operation with the United States in nuclear science. Before the 1955 conference, McIntosh had written to Hamilton saying that the high-level attention being given by the United States to the Atoms for Peace conference was a political fact that New Zealand must take into account. Whether or not Hamilton took any notice of this is unclear; he seems to have consistently recognised the need for nuclear power at some time in the future, but not for progress in the field of nuclear science research. In his report on the conference, Hamilton dismissed White's argument as sounding 'like a variant of "keeping up with the Joneses" ', adding that 'surely the influence of a nation depends on the standard of living of her people, in the broadest sense, and the contribution of ideas she can make in world affairs, not on whether she is doing atomic research irrespective of whether or not it is germane to her problems'. Hamilton advised that New Zealand should not undertake research in nuclear physics and technology, but rather should begin training 'a few bright young engineers-cum-physicists in general reactor design and operation in order to keep abreast of developments . . . and to assist in deciding what type of nuclear plant New Zealand should buy and when would be the appropriate moment to enter the market.'[16] Hamilton's antipathy towards White's attitude was reciprocated: Templeton has uncovered a personal letter sent in addition to his formal report in which White described Hamilton as 'hopeless as a delegation leader' and observed that he had clearly 'made up his mind in advance that New Zealand should not interest itself in atomic research'.[17]

Bill Latta, the State Hydro-electric Department's chief engineer, was perhaps the person best placed to judge New Zealand's need for a nuclear reactor for electricity generation. In his report on the conference, Latta agreed that there was 'no pressing necessity to construct atomic power stations in the immediate future' and said that it would be better to allow experimental nuclear power stations overseas to advance further before New Zealand committed to 'expending a great deal of capital on plant which might be obsolete before it went into service'.[18]

These New Zealand scientists and engineers were not going to be pushed into accepting a reactor — which, while a gift in part, would have cost a great deal of investment in terms of money and people power — just because a global superpower wanted them to. As well as refusing to be swayed by the American offers, New Zealand was also reluctant to enter into any sort of co-operation with Australia. At the conference, members of the Australian delegation made an informal approach to the New Zealanders regarding a scheme for Australian–New Zealand partnership in an atomic research project. While Philip Baxter, head of the Australian Atomic Energy Commission, and chief scientist Charles Watson-Munro, were keen for New Zealand and Australia to pool their efforts in nuclear science, the DSIR, under Hamilton, was not convinced. Baxter subsequently attributed New Zealand's failure to advance in the field of nuclear sciences to Hamilton's agricultural bias and failure to recognise the possibilities of the new field.

CABINET COMMITTEE ON ATOMIC ENERGY

Meanwhile, in the Department of External Affairs, Paul Cotton, a young graduate working in the Specialised Agencies Division, proposed that a

Cabinet committee be set up to respond to questions regarding the American offers of assistance and Australian offers of co-operation, and to study the general implications for New Zealand of the increasing development of peaceful uses of atomic energy. Acting on Cotton's recommendation, Cabinet set up a Cabinet Committee on Atomic Energy to study the reports of the delegation to the Atoms for Peace conference. At its first meeting, on 13 February 1956, the Cabinet committee established a committee of the permanent heads of the departments with an interest in the development of the peaceful uses of atomic energy. The Permanent Heads Committee on Atomic Energy was asked to report and make recommendations on: the question of a bilateral agreement with the United States; the American 'half-price' reactor offer; and participation in the Australian Atomic Energy Commission's research programme. They were also asked to prepare a draft statement of New Zealand's policy on the development of the peaceful uses of atomic energy.

Hamilton, who chaired the Permanent Heads Committee on Atomic Energy, drafted a policy statement, stating that New Zealand had no intention of setting up a research reactor in the near future and, provided the proposed Cook Strait submarine cable project — a plan to transport electricity from the South Island to the North Island — was feasible, would have no need for nuclear power for 30 to 40 years. Hamilton's draft was at odds with the Cabinet committee's enthusiasm for the United States' offer of a bilateral agreement and a half-price reactor; Cotton found the draft 'most unsatisfactory', and subsequently prepared a new draft of the report, that moderated Hamilton's views, for presentation at the first meeting of the Permanent Heads Committee on Atomic Energy.[19]

The Permanent Heads Committee subsequently recommended to Cabinet that the bilateral agreement with the United States be concluded as soon as possible. They also advised against accepting an offer of co-operation with Australia, believing that an atomic research partnership with the Australian Atomic Energy Commission would be costly, with no apparent benefit to New Zealand. In regard to the United States' offer of a half-price reactor, the Committee recommended that the immediate priority was to purchase an accelerator, but also recommended approval for the installation of a research reactor to be in operation in 'approximately three years' time'.[20] More significantly, the Committee noted that 'New Zealand cannot keep abreast of developments in nuclear science by merely seconding officers to work in overseas organisations such as Harwell in the United Kingdom or Chalk River in Canada. Officers seconded to these establishments will not return if there are no facilities in New Zealand in this field to enable them to pursue the advanced work for which they have been trained' and recommended the immediate establishment of an institute of nuclear sciences.[21] It advised that the new institute should incorporate the isotopes division of the Dominion Physical Laboratory, and be established as a branch of the DSIR, with an advisory committee representing the University of New Zealand, the DSIR and other interested parties.

At about the same time, the Council of Scientific and Industrial Research (CSIR) set up a small technical subcommittee to report on the future part New Zealand might play in atomic energy research and application. The CSIR subcommittee — which comprised Marsden, Hamilton, George Currie, Vice-Chancellor of the University of New Zealand, and Darcy Walker from Victoria University College — made recommendations to the CSIR, who adopted their report with minor modifications and conveyed it to the minister in a memo in June 1956. In line with the recommendations of the Permanent Heads Committee, the CSIR also recommended the establishment of an institute of nuclear sciences as a branch of the DSIR, with the immediate purchase of an accelerator, and a research reactor planned for two to three years' time. While similar to the Permanent Heads Committee's report, the CSIR report differed on two key points: it gave the need to purchase a research reactor much less emphasis, and it omitted the proposal to take advantage of the American offer of a half-price reactor.

Arguments that a bilateral agreement with the United States might hinder New Zealand's relationship with the United Kingdom had now dissipated: after New Zealand's original rejection of the United States' offer, the United Kingdom had told New Zealand that they would welcome an agreement between New Zealand and the United States, partly because it would make the exchange of information between New Zealand and the United Kingdom easier. By this time the United Kingdom may have been recognising the American superiority in nuclear technology and could see advantages to New Zealand having access to this technology.

Following the recommendations of the Permanent Heads Committee and the CSIR subcommittee, Cabinet decided to sign the bilateral agreement with the United States, while deferring consideration of the remaining proposals. The bilateral agreement between New Zealand and the United

States was signed on 13 June 1956. The agreement allowed for the exchange of information regarding the design, construction and operation of research reactors, and allowed for the lease of up to 6 kilograms of enriched uranium for use as reactor fuel. The United States signed a similar agreement with Iran in March 1957, which also included leasing Iran enriched uranium. It is ironic to learn that it was the United States who encouraged Iran's move towards nuclear technology: in 2002, United States President George W. Bush named Iran as one of three countries in an 'axis of evil' responsible for harbouring terrorists and seeking to create weapons of mass destruction, including nuclear weapons.

Although the bilateral agreement with the United States established a formal relationship between New Zealand and the United States on nuclear matters, the scientific relationship with the United Kingdom in the area of nuclear reactor and related technology — forged during the Second World War — was still strong, as was New Zealand's relationship with the United Kingdom as purchaser of New Zealand exports. New Zealand had an agreement to send up to three men a year to the Atomic Energy Research Establishment, and in 1959 would sign a formal agreement with the UKAEA regarding New Zealand's uranium resources.

While Cabinet was considering the full reports on the future of nuclear science in New Zealand, opposition to their recommendations was growing. The universities were concerned that locating the proposed institute of nuclear sciences anywhere other than a university campus would hinder opportunities for co-operation between the DSIR and the universities on nuclear science projects. Discussions on the subject of nuclear research revealed clear differences of opinion as to the value and urgency for New Zealand of an expanded nuclear research programme. In an August 1956 paper outlining the issues to his minister, the Secretary of External Affairs noted that the department's major concern in the matter was New Zealand's relationship with the United States, adding that New Zealand's failure to take advantage of the United States' offer of $350,000 towards the cost of a reactor 'may be difficult for the United States authorities to understand' and that the matter was 'causing some embarrassment in our relations with the Americans'.[22] Within the DSIR, the enthusiasm of individual nuclear scientists was not matched in the senior administration, where the proposed investment in research and training in nuclear sciences was seen only as a financial burden.

Despite the DSIR's seeming ambivalence on the issue, Cabinet decided to accept the United States' offer as part of a suite of decisions regarding New Zealand's nuclear future. The March 1957 Cabinet decision stated that New Zealand's policy in regard to research into and development of atomic energy should include:

(i)   opportunity for New Zealand scientists to keep reasonably abreast of developments in the uses of atomic energy overseas

(ii)   secondment of a small number of departmental officers for study at important nuclear stations in other countries

(iii)   purchase of an accelerator as a first step in the implementation of this policy

(iv)   the setting up of a suitable committee or institute to deal with the question of location, etc. when the purchase of an accelerator is authorised

(v)   an approach to the United States Government with a view to accepting their offer to pay a portion of the cost of a suitable reactor.[23]

With regard to the issue of an institute of nuclear sciences, representatives of the University of New Zealand also put forward their views that any such institute should be autonomous, rather than associated with a government department like the DSIR, and called for overseas experts to be brought to New Zealand to advise on the development of nuclear science in New Zealand.

At the request of the New Zealand Government, a six-man American nuclear advisory mission made a ten-day tour of New Zealand in March 1958, visiting and lecturing at various places. The team, led by the USAEC's Richard Kirk, visited with the intent to advise on the establishment of an atomic energy programme in New Zealand, specifically on the running of a nuclear institute, the siting and operation of a research reactor, the health and radiation problems involved in running a nuclear research institute, and the use of isotopes in agriculture, medicine and industry.

At the invitation of the University of New Zealand and the New Zealand Government, Philip Baxter, chairman of the Australian Atomic Energy Commission, and physicist Leslie Martin, also submitted a report on nuclear sciences in New Zealand in 1958. The university had pushed for the reports, but they were not in the university's favour. After the American and Australian reports were received, the university's argument was overruled and the decision was made to establish an institute of nuclear sciences as a branch of the DSIR. The plan was slow to be implemented, however, and the Government was criticised for the delays in getting the institute set up and ordering equipment. In January 1958, the *Otago Daily Times* editorialised that 'the atomic age is getting into its stride and New Zealand is already lagging several years behind'.[24]

ATOMS FOR PEACE CONFERENCE 1958

The Second International Scientific Conference on the Peaceful Uses of Atomic Energy was held under the auspices of the new United Nations International Atomic Energy Agency (the IAEA) in Geneva in September 1958. With more than 6000 participants, it was the largest international conference ever held.

The IAEA had been set up in 1957 as the 'Atoms for Peace' organisation proposed in Eisenhower's 1953 speech. In the words of its statute, its purpose was 'to accelerate and enlarge the contribution of atomic energy to peace, health and prosperity throughout the world'.[25] While its initial focus was on the peaceful use of atomic energy, the IAEA's work that came to be most relevant to New Zealand scientists was in the setting of international standards and regulatory procedures with regard to health and safety, which could be applied to the use of radioisotopes in industry, medicine and agriculture as well as atomic energy.

New Zealand was represented at the 1958 Atoms for Peace conference by Hamilton and Rafter from the DSIR, Marsden representing the NZAEC, Darcy Walker from Victoria University College, Hugh Parton, from the University of Otago, and John Scott of the New Zealand High Commission in London.

At the conference, Rafter presented the discovery of a link between atmospheric nuclear testing and raised levels of radiocarbon in the atmosphere, which he and Fergusson had first published in *The New Zealand Journal of Science & Technology* under the title 'The Atom Bomb Effect'. He and Fergusson had determined that the nuclear weapons tests of the 1950s had doubled the normal amount of atmospheric carbon-14 in the northern hemisphere, and increased southern hemisphere levels by 60 per cent. Since radiocarbon dating relies on comparing radiocarbon:carbon

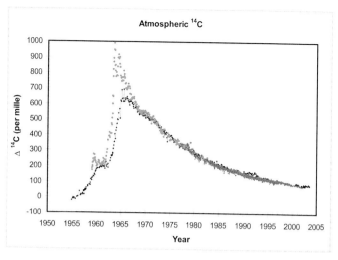

Atmospheric carbon-14 levels measured by Rafter and Fergusson from 1955 showed an increase in levels of carbon-14, which they attributed to atmospheric nuclear bomb testing. This graph depicts a clear spike in carbon-14 levels in the 1960s, when atmospheric testing was at its peak. The darker points are Wellington data, the lighter unconnected points are northern hemisphere data, with carbon-14 levels expressed as parts per million above normal levels. Courtesy Rodger Sparks, 'Radiocarbon Dating — New Zealand Beginnings', *New Zealand Science Review*, 61(2), 2004, pp. 39–41.

ratios in a sample with known ratios from specific time periods, this finding was highly significant, with the sudden spike of 'bomb' carbon allowing samples from the 1960s to be dated to within a year or two. Willard Libby was impressed by their paper, and contacted Rafter to ask if his team would participate in Project Sunshine, a USAEC project to measure environmental radioactivity resulting from nuclear bomb tests. As covered in Chapter Five, a network of monitoring stations was established in New Zealand with USAEC funding, with rainwater samples tested monthly for levels of strontium and other radioactive isotopes. In 1959, however, the network was passed on to the Department of Health's Dominion X-Ray and Radium Laboratory that was now responsible for monitoring fallout levels in New Zealand.

The trip was also an opportunity for New Zealand scientists to learn more about the northern hemisphere nuclear programmes, and to cement relationships with scientists in the United States and United Kingdom nuclear industries. After the conference, Hamilton and the other New Zealand delegates were guests of the UKAEA and visited the Atomic Energy Research Establishment at Harwell, the Calder Hall nuclear power station in Cumbria (which, when it opened in 1956, became the first nuclear power

In June 1958, Philip Holloway, the Minister of Scientific and Industrial Research, suggested that the new Institute of Nuclear Science's accelerator could be used to convert sawdust into poultry food. Cartoon by Gordon Minhinnick, *The New Zealand Herald*, 5 Jun. 1958, p. 12. Courtesy APN New Zealand.

station to produce electricity for commercial use), and Dounreay in the north of Scotland, where a team led by New Zealander Robert Hurst was experimenting with the first fast-breeder reactor. In these early years of nuclear technology, before the problems of disposal of a reactor's radioactive waste were acknowledged, a nuclear reactor that could produce more fissile material than it burned was considered an advantage. Hamilton then travelled to the United States, where he visited the USAEC's headquarters in Washington, Brookhaven National Laboratory at Long Island, and the nuclear laboratory at Los Alamos, which continued to be involved in the development of nuclear weapons.

Hamilton's tour of the British and American nuclear industries did little to sway his resolve that a nuclear reactor was not a priority for New Zealand science. In his *Dictionary of New Zealand Biography* profile of Hamilton, Ian Baumgart describes this agricultural scientist as being 'convinced that the country's economic future depended on scientific research that was focused on its resources and potentials'.[26] A nuclear reactor did not fit into this scheme.

In reporting on the 1958 conference, Hamilton repeated his conclusion from the end of the 1955 conference that New Zealand should not undertake research in nuclear physics or technology. He did recommend, however, that there was scope to extend the use of radioisotopes in industry and recommended the appointment of a full-time staff member to advise industry on how isotopes could be profitably employed. Illustrating the DSIR's agricultural focus, he also advised on an expansion of the use of radiation-induced mutations in the DSIR's plant-breeding programme. 'The cobalt source now at the Division of Nuclear Sciences is well suited for irradiation of plant material and no further additional facilities are required,' he wrote, although he did concede that 'the installation of an accelerator and, at a later date, a reactor, will permit a wider choice of ionising radiation'.[27] The possibilities of fusion as a power source were a feature of the 1958 conference, but Hamilton told *The Evening Post* that it was 'unlikely that electric power obtained by harnessing the hydrogen bomb — the fusion of deuterium and tritium — will be making a great contribution to the world's supply in the present century'.[28]

John Scott of the New Zealand High Commission in London also reported on the 1958 conference. Consistent with the support that the Department of External Affairs had been giving to American approaches to assist New Zealand with pursuing atomic energy for the future, Scott concluded his report by saying: 'Every country, including New Zealand, which aspires to high living standards, enlightened policies and practices for social betterment, and the maintenance of its position in the forefront of the progressive and developing countries of the world, must accept the challenge which atomic energy presents. It is a tool which will benefit greatly those who put use to it quickly.' While saying that the 1958 conference focused on developments in relation to fusion, Scott also reported on some of the latest predictions on the future use of isotopes, including 'the production of books using specially prepared paper that would last 20,000 years, the preservation of food, disinfestation of grain, and the treatment of inoperable cancer'.[29]

In New Zealand there were other unusual predictions about the possibilities of nuclear technology for the future. In announcing the decision to set up an institute of nuclear sciences and purchase an accelerator, Philip Holloway, the Labour Minister of Scientific and Industrial Research, talked about the 'miraculous possibilities' in the peaceful use of nuclear energy. As well as outlining applications of nuclear science to medicine, industry and agriculture, he said one possibility was that the new accelerator could process

1000 tons of sawdust a year into poultry food worth £20,000.[30] Newspapers dutifully reported his claims — how irradiation could turn sawdust into food was not explained — but at least one publication saw the absurdity of the statement and published a cartoon making fun of it.

INSTITUTE OF NUCLEAR SCIENCES

The establishment of the Institute of Nuclear Sciences helped to give status to the nuclear scientists already working within the DSIR, and gave New Zealand a second organisation — alongside the Department of Health's Dominion X-Ray and Radium Laboratory — focused on nuclear and radiation science. In 1958, as a step towards the establishment of the Institute of Nuclear Sciences, the DSIR's nuclear scientists were given the status of working under a separate DSIR Division of Nuclear Sciences with Rafter as director. As part of the decision that established the institute,

The DSIR's Athol Rafter pioneered radiocarbon dating in New Zealand, and with Gordon Fergusson perfected a more reliable method of radiocarbon dating using carbon dioxide gas rather than solid carbon. Rafter went on to lead the DSIR's Institute of Nuclear Sciences from 1959 until his retirement in 1978. Courtesy GNS Science.

Cabinet approved funding for the purchase of land, an accelerator, and the building of an administration and laboratory block. In response to the universities' concerns, the Institute of Nuclear Sciences was required to make provision for the use of facilities and laboratories by the universities and other government departments.

While Holloway talked about the 'miraculous possibilities' of the application of nuclear science to New Zealand's future, Hamilton was less enthusiastic. The most positive thing that Hamilton said about the DSIR's latest research unit was that the new Institute of Nuclear Sciences would help existing work in nuclear science by 'providing better equipment and facilities'.[31]

While the new institute would focus on benign scientific activities like radiocarbon dating and the use of isotopes for environment and industrial monitoring, there was some public apprehension about New Zealand's involvement in nuclear sciences. While rhetoric about atomic energy being the promise of the future was high in the 1950s, the reality in New Zealand was that the greatest public awareness of things nuclear was the American and British bomb testing in the Pacific that had ceased only when a temporary moratorium had been reached the previous year. In a 1959 address to a Nuclear Sciences Symposium in Wellington, Holloway said that the new institute must:

> enable our young men and women to become so used to the word 'nuclear' science, and so used to its practical application that they themselves will have confidence, and inspire confidence into the community. We must persuade the people that nuclear science is not necessarily a fearful thing; that reactors, accelerators, and all other machines used in its development, need not cause apprehension. The Press of the country must present the peaceful development of nuclear science clearly so that the public will accept it as a natural development and not something to be dreaded.[32]

At the same time as approving the Institute of Nuclear Sciences, Cabinet established the New Zealand Atomic Energy Committee (NZAEC), initially as an advisory committee for formulating the policy of the new Institute of Nuclear Sciences. One of the Committee members was Sir Ernest Marsden. In awarding Marsden a knighthood, New Zealand was acknowledging his work on the 1909 gold-foil experiment: while Marsden had made an important contribution to New Zealand science by leading the DSIR for two decades, it was after the Minister of Health spoke in Parliament about Marsden's role, with Rutherford, in the birth of nuclear physics that he was recommended for a knighthood for his services to science. The responsibilities of the NZAEC were to: advise on the organisation and administration of the Institute of Nuclear Sciences; make recommendations regarding the programme of research work at the institute and the funds required; advise on the co-ordination of New Zealand's activities in atomic affairs; encourage the publication and dissemination of the results of nuclear research; and advise on other matters relating to nuclear sciences.

The first challenge was to find a site for the new institute. The site had to be in the Wellington area and close to existing DSIR services, have adequate rock foundations for heavy equipment, and be at a safe enough distance from the public for the operation of a nuclear reactor. In 1959, Rafter was appointed inaugural director of the DSIR's Institute of Nuclear Sciences, and the first major piece of equipment — a 3-million-volt Van de Graaff accelerator — was ordered at a cost of £84,000. A hillside site at Gracefield, opposite the existing DSIR campus, was selected and the site developed. Cabinet finally, in 1960, approved the purchase of a nuclear reactor from the United States. But when they learned that it was no longer United States policy to give the half-price reactor subsidy to developed countries like New Zealand, the plans for acquiring a nuclear reactor were once again deferred.

Progress on construction of the building was slow, and when the Van de Graaff accelerator arrived in New Zealand in July 1961 it had to go into storage. When, by December 1961, there had been little further progress, the NZAEC considered the situation so serious that they approached the prime minister and members of Cabinet to outline the delays in the programme and seek continued assurance of support. The NZAEC addressed Cabinet on 5 March 1962, arguing the ways in which nuclear science was supporting New Zealand agriculture, forestry and environmental science, and making specific mention of: attempts to irradiate apple seeds to prevent germination during cold storage; giving Forestry Research Institute quick methods for determining variations in timber density; accurate and easy methods of gauging river flows; radiocarbon dating; and checking fallout from bomb tests.

The Institute of Nuclear Sciences building was finally completed and the accelerator assembled by 1966. Unfortunately, the magnet on the Van

This Institute of Nuclear Sciences building, on the hillside above Gracefield in Lower Hutt, had thick concrete walls to shield the high radiation levels sometimes generated by the 3 MW Van de Graaff accelerator. From left: Athol Rafter, first head of the Institute of Nuclear Sciences; Maurice Timbs, from the Australian Atomic Energy Commission; and Gordon Robb and Jim O'Leary from the New Zealand Atomic Energy Committee. Courtesy GNS Science.

de Graaff accelerator was damaged and the warranty had expired while it was in storage. Nonetheless, once repaired, the accelerator soon became an important tool for nuclear science research and environmental and industrial monitoring and was the institute's most significant piece of equipment until a more powerful tandem accelerator replaced it in 1986.

US NUCLEAR EQUIPMENT GRANT AND NEW ZEALAND'S FIRST NUCLEAR REACTOR

New Zealand science was as under-funded in the 1950s as it is today, and free laboratory equipment was always going to be well received by the DSIR and the universities, so when the USAEC offered gifts to support research in nuclear science, they were accepted with no apparent suspicions of ulterior motives on the part of the Americans.

The first proposal for a nuclear reactor in New Zealand had come from Marsden and Watson-Munro in 1947. Their plan was for an Australasian low-energy pile, which they believed would have 'defence significance'. This proposal never came to fruition, and later plans for a research reactor at the Institute of Nuclear Sciences were continuously deferred. In 1961, however, New Zealand did get a nuclear reactor, although rather than being associated with the DSIR it was installed in Canterbury University's engineering school.

After the USAEC visit in 1958, the American ambassador had suggested to Holloway that certain items of equipment that the visitors considered would be of immediate use in New Zealand institutions for research and training might be made available by the United States authorities under arrangements allowed for in the 1956 bilateral agreement. In March 1959, the Labour Prime Minister Walter Nash formally replied to the American ambassador to express New Zealand's interest in the proposal, attaching a list of equipment requested by Auckland and Canterbury universities and the Institute of Nuclear Sciences. The physics and chemistry departments of the University of Auckland asked for laboratory equipment worth US$65,672, the nuclear engineering laboratory of the Department of Electrical Engineering of the University of Canterbury requested funding for a sub-critical research reactor and ancillary equipment worth US$130,000, and the Institute of Nuclear Sciences requested a mass spectrometer and other equipment worth US$102,280 — a total request of US$297,952 (worth more than US$2 million in 2012 terms).

So while the Institute of Nuclear Sciences was still expecting to gain a nuclear reactor at some stage in the future, the University of Canterbury was the site of New Zealand's first — and only — nuclear reactor. Nuclear power was seen as inevitable for future power generation, and this was a valuable opportunity to train nuclear engineers in New Zealand. The sub-critical reactor at the University of Canterbury arrived in the electrical engineering department in 1961, under the care of professor of electrical engineering Norm MacElwee, who a few years earlier was reported as saying that 'it does not appear that nuclear power would have any advantages over hydro-electric power in the near future'.[33] The sub-critical reactor, by definition, had no critical mass of fuel to produce a chain reaction; its operation depended on neutrons being continuously added from an outside source.

The reactor used 2.5 tons of natural uranium as a fuel, a solid mixture of plutonium and beryllium as a neutron source, and tap water as a moderator. However, a reading of the Dominion X-Ray and Radium Laboratory files suggests that the engineering department accepted the gift without knowing much about what it entailed. In order to approve the importation of radioactive material, required under the Radioactive Substances Act 1949, the Dominion X-Ray and Radium Laboratory needed specific details of the reactor's fuel and source. The laboratory had to write to the USAEC, the donator of the gift, for details, confessing that 'no-one in New Zealand has any information on the physical properties of the uranium — in fact we are not even sure if it is uranium metal or uranium oxide, or whether U235 has been extracted'.[34]

While a nuclear reactor was generally considered to be a positive thing — a functioning piece of machinery that would help to train nuclear engineers for New Zealand's future — there was, by now, a degree of apprehension about the radioactive materials needed to fuel the reactor. The secretary of the NZAEC, Jim O'Leary, described in a letter to the Dominion X-Ray and Radium Laboratory that there could be 'a great deal of loose talk and emotion regarding the danger involved in such material'.[35] As predicted, the October 1961 delivery of the plutonium/beryllium neutron sources, which arrived by ship to Lyttelton, was considered newsworthy — although in today's light it is remarkable what little excitement a shipment of plutonium aroused. *The Press* reported that the 'plutonium' label in the ship's manifest caused 'a stir' amongst the crew when the cargo came on board, but they were reassured by shipping authorities, who in turn had been assured by the Dominion X-Ray and Radium Laboratory, that the cargo was safe.[36] This shipment, the first shipment of plutonium and the most powerful neutron source ever to arrive in New Zealand, consisted of three small cylinders, each about 2.5 centimetres in diameter and 4 centimetres long, sealed inside a large drum filled with paraffin wax to absorb any neutrons emitted from the cylinders.

The University of Canterbury's sub-critical nuclear reactor was soon operational. From 1964 onwards, the School of Engineering prospectuses advised that all third professional year electrical engineering students would attend a short lecture course on the electrical aspects of nuclear engineering. A later elective course, Advanced Electrical Engineering, focused almost entirely on nuclear engineering and used the reactor for laboratory

The Institute of Nuclear Sciences put on pikelets and a cream sponge to welcome American Ambassador Anthony Akers, and members of the NZAEC, in 1962. Seated at the table, from front left, are: Akers, Gordon Robb, Sir Ernest Marsden, Ian Dick and Dick Willett. At front right is Athol Rafter. Courtesy GNS Science.

demonstrations and experiments, although for some reason this was not made explicit in the name of the course. It is also surprising that neither the engineering school annual prospectus, nor *Student Engineer*, an annual booklet published by the Engineering Society of the School of Engineering, mentioned the sub-critical reactor, which as the only nuclear reactor ever to operate in New Zealand would surely have been a drawcard for the school.

But there was nothing secret about the reactor. Richard Duke, an electrical engineering student who took the nuclear engineering course in 1973, recalls the reactor being installed in a room with internal windows, through which the general student population and visitors could observe its daily use. The reactor seemed to draw no opposition, and at the School of Engineering's annual open days 'there were always long queues of people waiting to climb the steps to peer into the reactor tank and see the rods,' recalls Duke.[37] In 1981, by which time it was clear that New Zealand had

NEW ZEALAND CONTRIBUTIONS TO INTERNATIONAL ATOMIC ENERGY AGENCY (IAEA)

YEAR	FIXED CONTRIBUTION (US$)	NUCLEAR SECURITY FUND (NZ$)
1958	16,356	
1959	20,378	
1960	22,788	
1961	24,055	
1962	25,816	
1963	27,066	
1964	27,436	
1965	29,309	
1966	29,465	
1967	31,191	
1968	34,556	
1969	35,929	
1970	39,115	
1971	44,061	
1972	44,914	
1973	56,305	
1974	72,723	
1975	78,138	
1976	100,411	
1977	108,775	
1978	149,102	
1979	177,409	
1980	216,869	
1981	227,078	
1982	215,632	
1983	226,796	
1984	238,682	
1985	235,137	
1986	289,098	
1987	332,720	
1988	353,020	
1989	357,457	
1990	408,172	
1991	480,699	
1992	459,209	
1993	506,895	
1994	524,409	
1995	579,195	
1996	643,499	
1997	606,890	
1998	541,328	
1999	506,745	
2000	458,040	
2001	411,114	
2002	474,958	
2002/03	560,735	25,000
2003/04	670,126	20,000
2004/05	704,182	20,000
2005/06	699,922	25,000
2006/07	748,611	25,000
2007/08	995,446	–
2008/09	945,618	75,000
2009/10	1,084,185	40,000
2010/11	1,159,816	75,000
2011/12	1,190,401	75,000

Sources: Adapted from figures provided by New Zealand's Ministry of Foreign Affairs and Trade, Vienna and Wellington, by emails to author 2002, 2010, 2012. Figures from 1958 to 2002 were provided to the author in US$. Subsequent figures have been calculated from figures provided in a mix of €, US$ and NZ$, and are approximate.

no need for nuclear power, at least until after the end of the century, the University of Canterbury ceased offering the nuclear engineering course, closed down the nuclear engineering laboratory, and dismantled the reactor. The uranium in it went to the Institute of Nuclear Sciences, and the neutron sources containing the plutonium went to the university's physics department where it was used in research before being recalled to the United States after reaching its 30-year lifespan.

While the American gifts under the Atoms for Peace programme were of benefit to the New Zealand scientists and institutions whose laboratories they went to, it was not altruism on the part of the United States. The terms of the USAEC's gifts of research equipment to the University of Auckland's physics and radiochemistry laboratories were that the results of any research deriving from the use of equipment and materials would be provided to the United States. The equipment arrived in November 1960, and Ted Collins of the physics department wrote of the 'fever of excitement in the Chemistry Department as they open up their Xmas Box from Uncle Sam'.[38] When American Ambassador Anthony Akers visited the university's new radiochemistry and physics laboratories in October 1961, he expressed his view 'that your countrymen share the American dream of a world at peace, and it is with great pleasure that I participate here today in this programme reflective of the peaceful use of the atom'. A year later, the United States was testing more hydrogen bombs in the Pacific, despite Akers's assurance that it was 'the cherished hope of the American people that nuclear energy might be used only for peaceful purposes'.[39]

ATOMS FOR PEACE

While New Zealand scientists did not embrace the nuclear age with the enthusiasm that their American friends might have hoped, atomic energy was still part of New Zealand's vision of the future, and the United States nuclear submarine USS *Halibut* was greeted with awe and enthusiasm when it visited Auckland and Wellington in 1960.

The Institute of Nuclear Sciences was well established by the mid-1960s and was conducting original research as well as providing services to agriculture, industry and medicine. Director Athol Rafter continued to hope for the long-promised nuclear reactor, telling a visiting group from the

National Research Advisory Council in 1965 that a nuclear reactor would be 'of immense value to the nation and its scientists, and of particular value to industry'.[40]

By now, the International Atomic Energy Agency was working to encourage and support a predicted massive expansion in nuclear power generation internationally. New Zealand, as a member state, was making annual contributions to the IAEA's budget, which covered most of the regular functions of the agency, such as staff, conferences and some technical assistance. The size of New Zealand's contributions to the IAEA budget — more than US$1 million annually in recent years — can be seen as an example of how successful the nuclear industry has been in promoting a multilateral approach that has spread the costs over all countries, even New Zealand, which has never had a nuclear industry.

When the United States, the United Kingdom and the Soviet Union signed the Limited Test Ban Treaty in 1963, there was new hope that the world might be moving away from the endless development of new and more powerful nuclear weapons, and that nuclear technology would now be used for peaceful purposes rather than for weapons. But the early nuclear power stations in the United Kingdom, the Soviet Union and the United States — the first so-called 'peaceful uses' of nuclear technology — actually had the primary purpose of producing fissile material for use in nuclear weapons. But there was now a massive expansion in nuclear power stations for electricity generation even in countries with no nuclear weapons programmes.

New Zealand was about to face a new decision with regard to a nuclear future. In 1964, the New Zealand Electricity Department flagged that a nuclear power station would be needed to meet New Zealand's electricity demands by about 1977, and began the search for a suitable power station site and started training engineers for the task.

7

★

Nuclear decision
Plans for nuclear power

It is safe to say that within about half a century electricity from nuclear sources will be supplied to houses and small industries under much the same conditions as the present water supply.
— ERNEST MARSDEN, JUNE 1955[1]

... nations may in the future be rated as advanced or backward, developed or under-developed, according to their success in applying atomic energy to the solution of their problems.
— EXTERNAL AFFAIRS, 14 JUNE 1957[2]

In the 1960s, the south coast of the Kaipara Harbour, on the west coast north of Auckland, was considered an excellent place for New Zealand's first nuclear power station. It was close enough to the growing population of Auckland to make electricity transmission affordable, but remote enough to be considered safe in the event of a nuclear accident. Cooling water was available from streams in the Waitakere Ranges, and the harbour offered an outlet for wastewater from the plant. The original plan, for a nuclear power station with four 250-megawatt reactors, would meet most of Auckland's electricity demand.

In this *Auckland Star* cartoon, Prime Minister Sidney Holland, who was criticised for going slowly on nuclear power, is seen driving the 1911 hydroelectric model vehicle. Cartoon by Neil Lonsdale, *Auckland Star*, 27 Apr. 1956.

EARLY PLANS FOR NUCLEAR POWER

New Zealand first considered nuclear power as an energy source in the 1950s, when energy from the atom seemed poised to offer the world a safe, cheap, clean and almost limitless supply of electricity. The world's first nuclear power plant began operating in Obninsk, near Moscow, in 1954. Two years later, a British nuclear plant at Calder Hall in Cumbria began feeding power into the national grid. The young Queen Elizabeth, who opened the plant, announced that nuclear energy would be 'harnessed for the first time for the common good of our community'.[3] They were noble words, but the primary — and secret — purpose of the Calder Hall nuclear reactor, and many of the reactors that followed, was to produce weapons-grade plutonium. The generation of electricity, a secondary purpose of the reactors, was a way to help fund the weapons programme.

Internationally, the rhetoric for atomic energy was high. In his address to the 1955 Atoms for Peace conference, Indian nuclear physicist and president of the United Nations Conference on the Peaceful Use of Atomic Energy, Homi Bhabha, said 'for the full industrialization of the under-developed areas, for the continuation of our civilisation and its further development, atomic energy is not merely an aid: it is an absolute necessity. The acquisition by man of the knowledge of how to release and use atomic energy must be recognized as the third great epoch in human history.'[4] In the United States, the chairman of the USAEC spoke of an atomic future where electricity would be 'too cheap to meter'.[5] Australia's plans for nuclear power were already well in advance of New Zealand's when, in 1955, Charles Watson-Munro described atomic power as 'a coming force that would be comparable with the first onslaught of electricity on civilisation'.[6]

But New Zealand's interest in nuclear power was about more than just keeping up with the rest of the world. By the mid-1950s, electricity demand in New Zealand was rapidly outstripping supply — particularly in the North Island — despite the regular commissioning of new hydroelectric power stations. In a 1955 paper on the economics of nuclear power in New Zealand, Tony McWilliams described New Zealand as 'probably the only country in the world with a relatively high standard of living which has a continuing and serious power shortage'.[7] That year, in response to the need for a more systematic planning process, Cabinet set up a Combined Committee on the North Island Power Supply, initiating a system of annual planning reports. That same year, Geothermal Developments Ltd was formed to produce electricity and heavy water at Wairakei, uranium was found in Buller Gorge, and the New Zealand delegation returned from the first United Nations Atoms for Peace conference.

In his report on the Atoms for Peace conference, DSIR Secretary Bill Hamilton wrote that there were three main possibilities to solve the North Island's looming electricity problem: bringing hydro-power from the South Island by submarine cable; conventional thermal plants burning coal or oil; or nuclear power. Hamilton's characteristically pragmatic approach stated that the choice between the three must rest on their respective economic advantages, but he did point out the benefits of the submarine cable. Not only was hydroelectric power a renewable resource that should be fully developed before we turned to non-renewable sources like coal and uranium, he wrote, it was independent from overseas supply and would not be threatened in the event of war. Hamilton also saw delaying the introduction of nuclear power as advantageous, as we would then have the benefit of technological advances and a likely cheapening of the power source. Bill Latta, chief engineer for the State Hydro-electric Department, was in broad agreement with Hamilton — the idea of a Cook Strait cable had come from Latta in 1950 — concluding in his report on the conference that 'there is no pressing necessity to construct atomic power stations in the immediate future'.[8]

Not everyone agreed. Before leaving for Australia, Watson-Munro advised that unless geothermal steam was found to provide a plentiful and cost-efficient energy source, the North Island would be forced to rely on atomic power within the next ten years. He considered the country's hydroelectric potential close to exhaustion, and advised that the use of a submarine cable to bring power from the South Island to the North Island was too problematic.

By early 1956, the solution to New Zealand's ever-increasing demand for electricity was down to a choice between nuclear power and a Cook Strait submarine cable. While the DSIR and the State Hydro-electric Department favoured the Cook Strait cable, many individual scientists preferred the nuclear option, and roused popular support for it, too. Not only was nuclear power seen as being more reliable than hydroelectricity, because it was not dependent on the weather, it was also able to be sited close to where the power was needed and costs were expected to come down by the time New Zealand was ready to commission a nuclear power station.

Francis Farley, senior lecturer in physics at Auckland University College was, along with Hamilton and Latta, one of New Zealand's delegates to the 1955 Atoms for Peace conference. Critical of the cautious attitude of these government officials, he said that 'the real obstacle to nuclear power in New Zealand is the over-cautious play-safe attitude that is adopted in Wellington' and predicted that if the 'present increase in power demand continues, we might expect to have 10 nuclear power stations by 1975 to 1980'.[9] Farley's colleague Professor Percy Burbidge agreed, arguing that 'the system of generating power in one island and transmitting it largely to the other is to be deplored'.[10] In an article in the *Auckland Star*, Farley criticised the Minister of Works and Minister in Charge of the State Hydro-electric Department, Stan Goosman, and the head of the State Hydro-electric Department, Arthur Davenport, and put forward his view that atomic power would be cheaper than a Cook Strait cable.[11]

Marsden, always keen to have his voice heard on nuclear issues, supported the argument for nuclear power, and said that the people who were holding

"THERE ARE NONE SO BLIND ..."

Stan Goosman — Holland's Minister of Works and Minister in Charge of the State Hydro-electric Department
— is seen being blinded by the light of atomic power: an *Auckland Star* editorial the previous week (27 April
1956, p. 4) had described those who said atomic power stations were years away as 'blind to the fact that Britain
is proceeding with a full-scale nuclear power programme'. Goosman was thought to be in denial about the
benefits of nuclear power for New Zealand's future. Cartoon by Neil Lonsdale, *Auckland Star*, 4 May 1956, p. 2.

New Zealand back from the new technology were 'lazy-minded conservative
diehards who are afraid of change' and frightened that nuclear science
had become 'a malevolent, uncultured arbiter of our destiny instead of the
traditional servant of the industrial revolution'.[12] Darcy Walker, Watson-
Munro's successor as professor of physics at Victoria University, and another
New Zealand delegate to the 1955 Atoms for Peace conference, wrote an
impassioned letter to the Department of External Affairs in which he referred
to nuclear science as 'the crowning achievement of man's research in the
physical sciences' and criticised New Zealand's 'lack of decision about atomic
energy'. Walker called for 'a bold and enterprising attitude towards atomic
energy' to help New Zealand retain its best brains and stimulate a new
outlook on the nation's industrial future. 'If nuclear science and engineering
is largely ignored,' Walker added, 'we forego the benefits of one of the most
powerful stimulating forces of modern science. If we have faith in the future
of New Zealand we cannot afford to do this.'[13]

The editor of the *Auckland Star* took the side of physicists like Farley, Burbidge and Walker. Under editorials headlined 'These scientists must not be ignored' and 'Only the best advice will do', the stance of government officials like Hamilton and Latta was criticised, with the comment that 'at a time of electricity crisis, when the whole country is paying for the mistakes of the past, the Government remains stubbornly deaf to the advice of men in a position to give the best possible advice on the practical economics of a nuclear power station'. The May 1956 editorial added that the 'people are in no mood to tolerate further short-sightedness over power planning. They see atomic energy as a probable, and perhaps the only effective, alternative to a future darkened by power shortages', and called for an investigation into the possibility of nuclear power.[14] The *Otago Daily Times* added to the argument against the submarine cable, describing Cook Strait as 'a region notoriously disturbed by strong currents and seismic shocks in which the seabed contains fissures of considerable width'.[15] It wasn't just the mainstream newspapers advocating nuclear power. The left-wing newspaper *The People's Voice* concluded a piece reporting Farley and Burbidge's arguments by saying: 'It is in the interests of workers, housewives, farmers and industrialists of New Zealand to have available a source of reliable power, and it is the responsibility of the government to provide it.'[16]

As the argument between the university physicists and the government officials continued, other newspaper headlines warned of looming power-cuts if Auckland consumers failed to meet voluntary savings targets: there was high public awareness of the need for a new power source. But the Government line about nuclear power remained cautious. Goosman warned New Zealanders not to harbour premature hopes that nuclear power would soon be brought into New Zealand, saying it would not be economically feasible for many years. Goosman challenged Burbidge's argument that atomic power would be cheaper than hydro-power and wondered 'on what ground, or by what authority, Professor Burbidge sets himself up as an authority on practical power supply economics'.[17] Hamilton remained circumspect, with newspapers reporting his views that nuclear power held no promise of cheap electricity and it would probably be 20 years before New Zealand had need for trained engineers capable of supervising a nuclear power station. The Ministry of Works agreed, with senior engineer Jack Ridley stating that while atomic power would be one of New Zealand's major sources of energy in the future, it would not be an economic proposition for New Zealand within the next 20 years.

The voices of overseas experts always carried weight in a domestic argument, and in May 1957, when Leonard Cronkhite, director of the United States Atomic Industrial Forum, visited New Zealand his opinion was widely reported. Cronkhite voiced his support for the Government stance, saying that New Zealand appeared to be the least in need of atomic power of any of the twelve countries he had visited on his world tour. Newspapers reported his views that New Zealand was 'fortunately situated with natural power resources, such as rainfall, fast-flowing rivers, and geothermal power resources. The country could probably use the money required by an atomic reactor to much better advantage by producing power by these means.'[18]

Opinions remained divided, though, with individual scientists tending to be the ones voicing their endorsement of nuclear power. As with the current climate-change 'debate', these vocal dissenters got more than their share of media coverage. DSIR physicist Tony McWilliams, after returning to New Zealand from Harwell (where he had been working on the fractional distillation of heavy water in preparation for the planned Wairakei heavy-water plant), pushed for the early use of nuclear power plants in New Zealand. In a widely reported 1957 address to the Wellington branch of the New Zealand Institute of Engineers, McWilliams said that 'nuclear power shows promise of satisfying completely the incremental power demand with considerably less strain on the economic resources of the country. It frees the power supply from the vagaries of the weather which will always plague hydro-electric power. Great advantage should stem from the use of nuclear power and its early introduction should be seriously considered.'[19] He added that nuclear power stations could be introduced to New Zealand as early as 1962. There were also some warning voices from within the universities, however, and as early as 1955 Professor Rastrick, of Canterbury University College's mechanical engineering department, warned that the problem of disposal of radioactive waste would have to be overcome before atomic power could be generated in New Zealand.

Today, the Labour Party is closely associated with introducing New Zealand's nuclear-free policy, but when it came to decisions about nuclear power in the 1950s they were more pro-nuclear than National. At the end of 1957, after a Labour Government led by Walter Nash was elected, the new Minister of Works and Electricity, Hugh Watt, responded to public calls for nuclear power with an announcement that the Government would launch an investigation into the possibility of nuclear power stations in New Zealand. While in opposition, the Labour Party had opposed the Cook Strait cable project, and Watt now said that the investigations into nuclear power would be more intense than under the previous Government and would be conducted with the view that electricity produced by nuclear means was a distinct possibility for the future.

The State Hydro-electric Department duly reported in 1958 that atomic energy was a 'promising source of power'. It continued, however, to say that New Zealand had 'natural sources which, at the moment and for some few years ahead, seem likely to provide power more economically and with less drain on overseas funds'.[20] It recommended that atomic energy be reconsidered in five years. The report continued by approving the construction of a new hydroelectric station at Benmore on the Waitaki River, initially to supply power to the South Island, and deferred a decision on linking the islands with a Cook Strait cable.

By 1961, after numerous technical issues had been resolved, the Government approved the scheme to link the North and South Island power systems, and a contract for manufacturing and laying the cables was placed. In 1965 a submarine high-voltage DC cable — only the third of its kind in the world — finally linked the North and South Islands. Most of the power from the Benmore station was carried across the cable to provide the growing population of the North Island with, at last, a plentiful and reliable source of electricity. But while the Cook Strait cable had won the toss-up between nuclear power and a link between the North and South Islands, it now seemed that both solutions would eventually be required.

The annual reports of the New Zealand Electricity Department (NZED)'s Planning Committee on Electric Power Development in New Zealand continued to project future electricity demand and detail plans for future power sources for New Zealand.* The 1964 report of the planning committee contained the first mention of nuclear power as a possible source of electricity for New Zealand. In this report, hydro, geothermal, natural gas, oil, coal and nuclear sources were all considered to meet New Zealand's future and rapidly escalating demand for power. In considering nuclear power as an

---

* The State Hydro-electric Department had become the New Zealand Electricity Department in 1958, reflecting the country's diversifying sources of electricity.

option, the committee noted that 'there is no doubt that this means of power generation must be introduced in New Zealand by about 1977'.[21] It also had the foresight to note that the cost of setting up such a plant could be £100 million or more, and recommended that a possible site for a nuclear reactor be selected, and a meteorological observing and recording station set up to provide details on local atmospheric conditions.

TRAINING NUCLEAR ENGINEERS

While there had been many arguments about the timing of the introduction of nuclear power to New Zealand, and the relative benefits of different energy sources, it was widely accepted, even amongst the cautious DSIR hierarchy, that nuclear power would be introduced at some stage in the future, and that scientists and engineers should be trained for this eventuality. As discussed in Chapter Six, in 1961 the University of Canterbury set up a nuclear engineering course, complete with a sub-critical nuclear reactor in the laboratory, to train engineers for the expected New Zealand industry.

The responsibilities of the New Zealand Atomic Energy Committee (NZAEC) expanded to include liaison between organisations and departments planning for the introduction of nuclear power, and in 1964 the committee set up siting and manpower subcommittees to study suitable sites for a nuclear power station and the manpower requirements for the operation of a power station. The NZAEC Manpower Subcommittee was chaired by Robin Williams (one of the young New Zealand scientists who had worked on the Manhattan Project during the Second World War) and included representatives from the NZED, the Ministry of Works, the DSIR, and the Institute of Nuclear Sciences.

The Manpower Subcommittee first reported in 1966, outlining recommendations for a training programme to enable the NZED and the Ministry of Works to train staff to work with consultants engaged in the siting, design, construction and start-up of a nuclear power station. A first stage — sending two senior engineers for a four-month nuclear engineering course at Lucas Heights in Australia — had already taken place. Henry Hitchcock of the NZED described the 1966 course, which he attended as a 54-year-old research engineer, as both interesting and traumatic. He recalled arriving at

The ANZAC team, at the end of their two-year stint, at the UKAEA Reactor Design Office at Risley. The New Zealand team includes Rob Aspden (in back row, smiling, third from left), Hector Jones (in front and to the right of Aspden) and Neil Fyfe (back row, obscured). Three other members of the New Zealand team are absent from the photograph. Courtesy Rob Aspden.

the course 'equipped with stage one physics and stage one chemistry which I'd passed in 1930, which is two years before the neutron was discovered. Nuclear physics was a big jump . . . I was dragged from the beginning of the twentieth century to the second half of the twentieth century by the scruff of my neck.'[22]

A programme sending more engineers — nuclear, electrical, mechanical and civil — as well as a reactor physicist and a health physicist was to follow, with men travelling to the Australian School of Nuclear Technology at Lucas Heights, the University of New South Wales in Sydney, and Imperial College London for intensive nuclear engineering education. The subcommittee rejected the idea of expanding training facilities in New Zealand, but was operating under the expectation that training would eventually be available at the proposed research reactor at the Institute of Nuclear Sciences.

One large team spent two years training in the United Kingdom. In May and August 1967, six New Zealanders — five engineers and a radiation physicist — travelled to the UKAEA Reactor Design headquarters at Risley, Lancashire, where they worked on a joint British and Australian project to adapt the enriched-uranium fuelled Steam Generating Heavy Water Reactor into a reactor that could be fuelled by natural — rather than enriched — uranium. Hector Jones of the NZED led the New Zealand group, which included men from the Ministry of Works, the NZED, and the Department of Health's National Radiation Laboratory. The New Zealanders' main objective in this project was to become competent in the technology associated with a water-moderated nuclear reactor, so as to form a team of engineers capable of working with overseas consultants when it came time for New Zealand to commission a nuclear reactor.

The project was completed in 1969 — the same year the Maui gas field was discovered in offshore Taranaki. After the course, three members of the New Zealand team stayed on in London to study reactor safety assessment and licensing with the Ministry of Power's Inspectorate of Nuclear Installations. They also visited Germany's and Canada's heavy-water reactors. In their report to the NZED, they concluded that even if the discovery of natural gas deferred the planned New Zealand nuclear programme, by then global advances in operating nuclear power stations of the type suitable for New Zealand conditions would have built up to the point where reasonably developed and proven systems will be available. 'Such a delay should not mean any let-up in our efforts to prepare for the introduction of nuclear power stations into New Zealand,' they added. 'On the contrary, the "breathing space" should be utilised to continue to build a pool of engineers and scientists trained in the various technologies and know-how involved, and who will be capable of forming a solid foundation on which successful nuclear power generation in New Zealand must ultimately rest.'[23]

Despite increased uncertainty about the future of nuclear power for New Zealand caused by the Maui gas discovery — which had the potential to solve New Zealand's electricity needs for the medium term — the NZED, the Ministry of Works, the Department of Health, and the Commission for the Environment continued to send staff for training at Lucas Heights and other international nuclear technology schools. By the beginning of 1976, 26 New Zealanders had undergone some form of overseas training in nuclear technology.

FINDING A SUITABLE SITE

Auckland's rapidly expanding population was creating a big increase in electricity demand. In his 1964 report on nuclear power generation, Eric Mackenzie, general manager of the NZED, recommended that serious consideration be given to locating a nuclear power station north of Auckland city, given 'the very difficult transmission problems concerned with delivering power to the Auckland area'.[24] In December 1964, the Siting Subcommittee of the NZAEC met for the first time and, responding to recommendations tabled by the NZED, selected two sites on the southern shores of the Kaipara Harbour — at Oyster Point and South Head — as the preliminary preferred sites for New Zealand's first nuclear power station.

Philip Blakeley of the NZED led the Siting Subcommittee, which included representatives from the Ministry of Works, the Meteorological Service, the National Radiation Laboratory (Jim McCahon), and the Institute of Nuclear Sciences. In selecting a site, the subcommittee had to consider health and safety issues, engineering and transport requirements, and the availability of cooling water. While sites at South Head and Oyster Point were the focus of investigations, the subcommittee also considered a further eighteen sites between the Manukau and Kaipara Harbour entrances on the west coast — including a site at Bethells Beach and one between Piha and Karekare — and between the Whangaparaoa Peninsula and Pakiri on the east coast. Water for the proposed power station would come from streams in the Waitakere Ranges and in the hills north of Puhoi or by drilling to the water table.

In 1965, the Meteorological Office installed anemometers to measure wind speed at the two favoured sites on the Kaipara Harbour, and the NZED began measuring seawater temperatures. The New Zealand Navy conducted a hydrographic survey near the two sites, and the New Zealand Geological Survey conducted seismic surveys to investigate the underlying geology. These investigations confirmed by 1968 that the South Head site was a suitable location, and the focus of investigations shifted to Oyster Point, which had the advantage of being closer to Auckland.

Opposite: The DSIR Geological Survey mapped the coastal areas north of Auckland being considered to locate a nuclear power station and prepared this map in 1966. Sites 19 and 20, on the Kaipara Harbour, were considered the most promising sites, and more detailed geological, meteorological and oceanographic investigations followed. ABLP W4215, box 2, 3-2, pt 1, Archives New Zealand, Wellington, New Zealand.

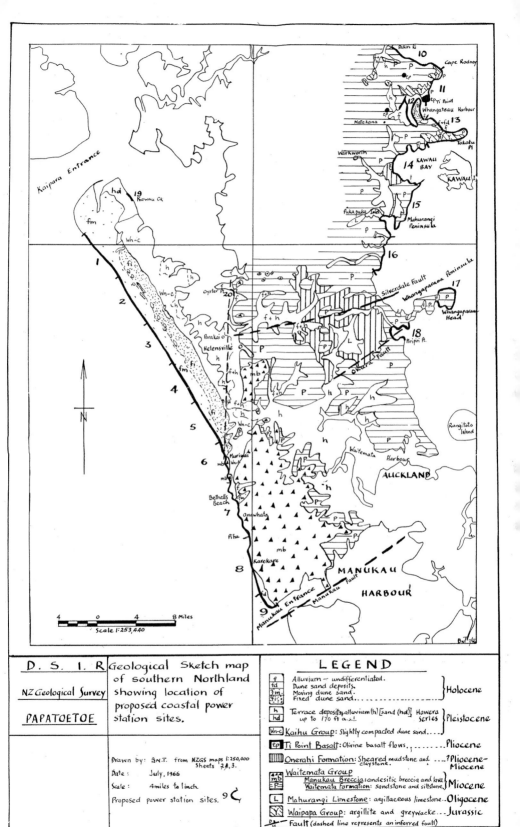

LEGEND

f	Alluvium — undifferentiated.	
fd	Dune sand deposits.	Holocene
fm	Moving dune sand.	
fi	Fixed dune sand.	
h / hd	Terrace deposits; alluvium (h) [sand (hd)] Hawera series up to 170 ft a.s.l.	Pleistocene
Wn~c	Kaihu Group: Slightly compacted dune sand.	
tp	Ti Point Basalt: Olivine basalt flows.	Pliocene
	Onerahi Formation: Sheared mudstone and claystone.	?Pliocene-Miocene
mb / P	Waitemata Group: Manukau Breccia: andesitic breccia and lava. Waitemata Formation: sandstone and siltstone.	Miocene
L	Mahurangi Limestone: argillaceous limestone.	Oligocene
Y	Waipapa Group: argillite and greywacke.	Jurassic
	Fault (dashed line represents an inferred fault)	

D. S. I. R

NZ Geological Survey

PAPATOETOE

Geological Sketch map of southern Northland showing location of proposed coastal power station sites.

Drawn by: B.N.T. from NZGS maps 1:250,000 Sheets 2A, 3.

Date: July, 1966

Scale: 4 miles to linch.

Proposed power station sites. 9

Scale 1:253,440

4 0 4 8 Miles

In considering the best site for a power station, priority was given to the safety of the human population in the event of an accident. One of the issues faced was the risk of earthquakes, and in 1971 the New Zealand Atomic Energy Committee set up a Working Group for Seismic Effects on Nuclear Installations to 'examine and report on problems relating to design requirements and the construction of nuclear reactor installations in New Zealand, resistant to seismic effects'.[25]

The impact of the day-to-day running of a nuclear power station on the land and the marine environment was given less consideration. A report was received, however, from the New Zealand Oceanographic Institute, which advised that the discharge of warm-water effluent from a power station into an enclosed shallow harbour like Kaipara, the largest harbour in Northland, would significantly raise the water temperature, which, as well as encouraging the breeding and growth of species already present, could encourage the establishment of exotic warm-water species arriving on ships. It also said that algal blooms and various destructive marine-boring organisms could be more likely in the warmer waters. In summing up, the Oceanographic Institute said that altering the temperature regime in the harbour could 'affect a large number of people and cause considerable public inconvenience and anger . . . because it would affect the facilities for recreation . . . and commercial fisheries'.[26]

Given that Auckland was the city most in need of a new power source, it is perhaps not surprising that academics from the University of Auckland led the call for nuclear power. In the tradition of Farley and Burbidge before him, Auckland university's chemistry professor, Allan Odell (who was involved in many nuclear science projects), in 1965 said that New Zealand would benefit greatly through co-operation with Australia on the introduction of nuclear power. *The Dominion* editorial called his suggestion 'eminently sound', pointing out that Australia had already amassed a fund of scientific and engineering knowledge about nuclear power that could help save New Zealand time and money.[27] By now Australia's research reactor at Lucas Heights had been operating for five years. In another four years, in 1969, plans would be announced for a nuclear power station to be built in Jervis Bay in New South Wales. Tenders for the reactor, which would have to be supplied from overseas, would be called for and $1.25 million spent on building access roads, houses for future employees, and water and power services.

COMMISSION DATE SET

While earlier reports identified that nuclear power would be needed by about 1977, it was not until 1968 that plans for a nuclear power station came within the NZED's ten-year planning period. The 1968 planning report recommended that a 250-megawatt reactor start operation in 1977, with three similar units following at yearly intervals, to build the station up to 1000 megawatts of generating capacity. As well as observations made at the favoured Kaipara Harbour sites, consideration had been given to possible additional nuclear power station sites in the Firth of Thames, south of Auckland City, and at Baring Head, near Wellington. The 1968 report concluded, however, that the programme for the introduction of nuclear power could be significantly affected in the event of early large-scale discoveries of natural gas, which would have the advantage of being an entirely indigenous resource.

The 1969 report pushed the commissioning date for the first 250-megawatt nuclear power station forward one year, to 1978. Also in 1969, the British company of Preece, Cardew, and Rider were engaged to prepare a report studying the economics of nuclear power generation in relation to hydroelectric and thermal power generation. If nuclear power was found to be economically viable, they would then proceed to make a specific proposal suitable for New Zealand conditions.

In May 1969, *The Evening Post* suggested that natural uranium and heavy water would be the fuel and the moderator for New Zealand's first electricity-generating nuclear reactor, saying an advantage was that the fuel was low-cost and could be produced in New Zealand. The article also added that such a reactor 'produces a considerable quantity of plutonium for "fast-breeder" reactors', adding that 'because it is unlikely New Zealand would require plutonium until some years after the first nuclear station is established, the spent fuel would be stored away in underground tanks on the station site'.[28] It was not until the mid-1970s that plutonium produced by the reactors would be seen as a problem rather than an advantage, although even then environmentalists' main concerns were the possibility of some plutonium escaping into the atmosphere, or the threat of terrorism, rather than how to dispose of the highly radioactive waste.

Preece, Cardew, and Rider's final report on the Economics of Nuclear and Alternative Forms of Generation was received by the NZED in late 1970.

While the initial brief of the company was to confirm that nuclear power was the most economical way to meet growing demand in the North Island, and to recommend the most suitable type of reactor, the Maui gas discovery switched the focus of their investigations to economic comparisons between gas-fired, oil-fired, and nuclear-fuelled stations. Preece, Cardew, and Rider's report showed that the 'relative economic merits of the various types of generation considered depended on the price payable for fossil fuels, and the weighting given to overseas funds'.[29] The consultants recommended that if, after taking into account the above factors, nuclear-fuelled generation appeared to be the most attractive proposition, tenders should be called for a 1000-megawatt nuclear-fuelled station. Should the tenders confirm its economic advantage, construction of such a nuclear power station could proceed.

Indigenous fuel sources, however, were now proving more abundant than previously realised. By 1972, testing of the oil wells off the Taranaki coast had revealed large quantities of natural gas; more than enough to fire the 600-megawatt power station under construction at New Plymouth. At the same time, a reassessment of Waikato's coal reserves revealed enough coal in the Huntly area to support a 1000-megawatt thermal power station. In 1973, the New Zealand Government entered into a joint venture with an oil consortium — Shell BP Todd — to develop the Maui offshore gas field. In response, the NZED said that New Zealand could now defer the introduction of nuclear power stations until the late 1980s, or even later, should more indigenous energy resources be found. Even so, they made it clear that consideration should continue to be given to the resources required for a future nuclear power programme.

The early months of 1973 were very dry, and New Zealand's reliance on hydroelectricity meant more electricity restrictions and even occasional power-cuts and blackouts. Later in the year came the oil crisis, in which Middle Eastern nations set up an embargo on oil exports, leading to huge increases in oil prices. As well as limiting the possibility for thermal generation using oil, this crisis also highlighted the danger of relying on imported fuel supplies for electricity generation. Another dry summer followed. In response, the NZED's construction programme was advanced, the power planning committee pushed for sufficient capacity to meet dry-season demand, and the 1973 Huntly coal-fired station was approved.

The planning committee remained interested in nuclear power as an option for electricity generation, and in 1974 reported that 'other countries without the indigenous resources of New Zealand are embarking on a vigorous nuclear programme to meet their energy requirements' and so far as New Zealand was concerned 'plans for introducing nuclear power must be made'.[30] This was the first plan that adopted a fifteen-year planning period (as compared to the ten years used previously), meaning that a nuclear power station was back on the power plan. A North Island nuclear power station of two 600-megawatt reactors was scheduled for commissioning in October 1988, with the rider that the introduction of the station should be deferred if suitable alternative indigenous energy sources became available. The report noted that 'in any planning of nuclear power stations, safety will be a major consideration, and any proposals will have to comply with the exacting standards which would undoubtedly be laid down by whatever independent authority is responsible for the licensing and supervision of such installations'.[31] Appendix II contained information about radioactive waste disposal. It described how spent fuel from a New Zealand nuclear power station would be stored at the power station for up to three months, to allow the decay of short-lived fission products. It would then be loaded into specially constructed containers 'designed to remain intact in severe accident conditions' and transported to the reprocessing plant where the unburnt fuel would be recovered. It concluded that 'should a decision be made to install a nuclear station it will be necessary to conclude an agreement with an overseas reprocessing plant to treat the used fuel. The spent fuel would be transported to the plant, and the wastes treated and stored in that country along with the other wastes produced at the plant.'[32]

A CHANGING PUBLIC MOOD

In 1975, the Labour Government established a Fact Finding Group on Nuclear Power to report on the possible environmental consequences of a nuclear power programme in New Zealand. The group set up four expert working parties to study electricity supply and demand, nuclear power generation and reactor safety, health risks, and the siting of power stations.

In the two decades that New Zealand had been contemplating and discussing the introduction of nuclear power, the public mood had changed. Nuclear power plants had been generating electricity for domestic and industrial consumption for nearly 20 years, and were now in use in nearly

20 countries, including the United Kingdom, the United States, the Soviet Union, Canada, France and India. But because indigenous power sources had always been cheaper than the nuclear option, New Zealand had deferred its nuclear decision until the arguments for and against nuclear power were more sophisticated. Nuclear energy produced by fission had initially been promoted as a cheap, pollution-free, self-perpetuating source of energy. However, because of unexpectedly high costs, the unresolved problem of what to do with the radioactive waste it generated, and increased public opposition to nuclear energy, many plans for nuclear power plants had been shelved.

Public opposition had grown in part because of the industry's connections with weapons manufacturing (the British nuclear power plant at Calder Hall, for example, produced plutonium as well as generating electricity) and a series of nuclear accidents. A 1957 fire at Britain's first nuclear reactor at Windscale, for example, caused radioactive contamination of the surrounding countryside, and a reactor meltdown on a Soviet icebreaker in 1966 made international headlines. The images nuclear power conjured up were no longer so rosy. The *New Zealand Listener*'s Boyce Richardson in 1975 described nuclear reactors as now symbolising 'an impersonal future world of computers, robots, explosive violence and uncontrollable technology, rather than the cornucopia overflowing with goods and pleasures that they once promised'.[33]

Safety was an increasingly important concern. The nuclear power programmes of Europe, begun in the 1960s and boosted by the oil shock of 1973, were in a new phase, which *Listener* writer Geoff Chapple described as one of 'public challenge, planning delays and outright rejection'.[34] Anti-nuclear protestors had occupied nuclear reactor sites in Germany, France and the United States, staged street demonstrations in the United Kingdom and Australia, and caused informal moratoriums on the building of new reactors in West Germany and Japan. Opposition came from a number of grassroots sources — Australian scientist Francis Robotham described those fighting against the 'atomic juggernaut' as a 'strange and somewhat unique coalition of church groups, scientists, trade unionists, political parties and plain ordinary people'.[35]

The *New Zealand Science Review* in 1974 carried an editorial about the safety aspects of disposing of waste from a nuclear power station in New Zealand. The editor, J. G. Gregory, commented on Prime Minister Norman

Kirk's assurance that no decision on a nuclear power station for New Zealand would be considered 'until or unless there is an absolute assurance that it is possible to dispose of the poisonous and dangerous wastes of that power plant in complete safety'.[36] Gregory noted that 'even the United States Atomic Energy Commission did not claim to be absolutely sure of the safety of disposal and postulated the consequences of release of plutonium-239', and continued:

> More of our children would abort or be stillborn, more would be born with genetic defects, more would die in infancy, we ourselves might not live to see our grandchildren. All this would occur in an atmosphere of mental tension and ill health, social stress, and heightened civil unrest. . . . It is *not* an option to dump these wastes anywhere — in the sea, in salt mines, in space, in the Antarctic, anywhere.[37]

By now, claimed Elsie Locke in her book on the New Zealand peace movement, 'a profound distrust of nuclear energy in any form was widespread'.[38] The Campaign for Nuclear Disarmament had led a campaign opposing visits from nuclear-powered ships. Nuclear power was seen as potentially dangerous, and, since 1971, American nuclear-powered ships had not been allowed into New Zealand harbours unless the United States guaranteed liability for any damage to people or places in the event of a nuclear accident. Environmental movements had also emerged around the world, and in New Zealand the fight against the raising of Lake Manapouri to boost hydroelectric power generation had — in Boyce Richardson's words — 'moved forever the foundation on which official and public thinking about the environment of this country is based'.[39]

In 1974, the British Royal Commission on Environmental Pollution examined the impact of nuclear power on the environment at a time when the United Kingdom seemed on the threshold of a major commitment to nuclear power, in particular to the fast-breeder reactor. In its report published in 1976, the commission found that inadequate means for disposal of radioactive waste, as well as the risks of using and producing large quantities of plutonium, made the large-scale introduction of nuclear power 'irresponsible and morally wrong', adding that 'we should not rely for our energy supply on a process that produces such a hazardous substance as plutonium unless there is no reasonable alternative'.[40]

In March 1976, a coalition of environmental and anti-nuclear groups formed the Campaign for Non-Nuclear Futures to oppose the introduction of nuclear power to New Zealand, promoting instead alternative renewable energy options, like solar, wind and geothermal power. The main goal of the campaign was to collect half a million signatures in a petition against nuclear power. It was against this background that New Zealand considered whether or not to build a nuclear power station.

ROYAL COMMISSION OF INQUIRY INTO
NUCLEAR POWER GENERATION

The 1975 report of the Planning Committee on Electric Power Development, which had retained two 600-megawatt nuclear reactors, noted that 'if there is to be any chance of meeting a 1988 commissioning date it will be necessary for a decision to be made by 1977 on whether or not to introduce nuclear power'.[41] It was with this in mind that Robert Muldoon's National Government, which had come to power in 1975, set up a Royal Commission of Inquiry into Nuclear Power Generation: it had been one of the National Party's election promises that nuclear power would not be introduced to New Zealand 'until a public inquiry into all aspects of this source of energy has taken place'.[42]

The commissioners were announced in August 1976: the commission was to be led by Sir Thaddeus McCarthy, a retired judge and President of the Court of Appeal. The other commissioners were: Ian Blair, a plant pathologist with community involvement in protection against water pollution; Vivienne Boyd, vice-president of both the National Council of Women and the Wellington Association of Baptist Churches; Bruce Liley, professor of physics at the University of Waikato; and Lindsay Randerson, a businessman. There was some opposition to the appointment of Liley, whose research into the application of plasma physics to nuclear fusion reactors was cited by Campaign Half Million and some environmental groups, including Friends of the Earth, as evidence of his advocacy for nuclear power for electricity generation, and there were calls for his resignation. The Minister of Electricity, Eric Holland (son of the late Sidney Holland), defended Liley's appointment, arguing that they had appointed a truly impartial commission that would present credible and dispassionate recommendations.

The Campaign for Non-Nuclear Futures launched its Campaign Half Million to collect signatures opposing the introduction of nuclear power to New Zealand — something the New Zealand Electricity Department saw as inevitable. Meanwhile, many countries with nuclear power had moratoriums on new nuclear power stations and the global industry was in decline. Cartoon by Malcolm Walker, *Sunday News*, 4 Jul. 1976, p. 17.

Just a few days after the commissioners were announced, the Planning Committee on Electric Power Development in New Zealand tabled its 1976 report, in which a nuclear power station was now scheduled for 1990, with a decision on whether or not to proceed required by 1977.

Before the 1970s, the focus of anti-nuclear protests had been opposition to nuclear weapons testing, but the anti-nuclear movement was increasingly encompassing opposition to nuclear power and nuclear warships, and nuclear issues were coming to a wider public awareness, too. It is revealing to look at a single issue of *The Dominion*, from 27 August 1976. Page one carried two nuclear stories: one on the Planning Committee on Electric Power Development in New Zealand's plans for a nuclear power station by 1990, which began 'New Zealanders will probably have to accept nuclear power by 1990'; and one on the Civil Defence preparations for an upcoming visit by the nuclear-powered USS *Truxtun*. Under the Government's safety code for the entry of nuclear-powered ships to New Zealand ports, the national civil defence nerve centre under the Beehive was activated, and radiation monitoring devices were installed at 20 sites around Wellington Harbour. Page three of the newspaper carried a story about MPs debating and defeating a Bill calling for a South Pacific nuclear-weapon-free zone and the prohibition of nuclear weapons and reactors from New Zealand.

At about this time the links between nuclear power and nuclear weapons were also becoming clear, and not just to people in the peace movement. *The Dominion* editorial of 17 August 1976 had described international stability as being disturbed by countries acquiring nuclear weapon capability through projects meant for peace, describing the Eisenhower Atoms for Peace project, in which the United States provided many countries with nuclear reactors, fissile material, and training for their scientists, as being 'founded out of bad conscience'. The editorial continued to point out that 'the fuel from reactors . . . can produce a crude bomb' and 'the technology is freely available not only to governments anxious to develop a nuclear muscle but also to terrorist organisations'.[43]

Over the course of the Royal Commission of Inquiry into Nuclear Power Generation in New Zealand, the commission heard 141 submissions: from organisations potentially involved in the establishment of nuclear power stations in New Zealand, such as the NZED, the DSIR, and the Ministry of Works and Development; from environmental, peace, church and women's organisations; and from concerned or interested individuals. The commission members also travelled overseas to conduct interviews and visit energy installations in the United States, Canada, Europe and South Africa.

Most submissions were against the introduction of nuclear power, generally because of cost — other options were still assessed as being cheaper than nuclear — or for reasons of environmental safety and public health. Others criticised New Zealand's total energy consumption, urging a move towards a sustainable society promoting energy conservation and development of renewable resources. The inquiry report noted that submissions were, numerically, opposed to nuclear power, which respondents found 'an expensive, dangerous, imported technology'.[44]

Many of the individual and group submissions came from scientists, and it is interesting to note the change of attitude towards nuclear technology from the 1950s to the 1970s — especially where it is possible to track the changing opinions of individual scientists. In the 1950s, when the DSIR and the NZED were wary about the need for nuclear power, many individual scientists, including Percy Burbidge and Gordon Williams, were berating the Government's cautious attitude and talking enthusiastically to the media about the inevitability of nuclear power and the need to start training nuclear engineers immediately. By 1977, when he made a submission to the Royal Commission of Inquiry on Nuclear Power Generation, Burbidge was long retired from his position as professor of physics at the University of Auckland. In contrast to his 1950s calls favouring a nuclear power station over a Cook Strait cable, Burbidge now claimed that not only was there ample energy available from native sources of energy for the next 30 years, but 'the danger to our population and the potential damage to our industrial production are too great to justify the introduction of reactors for electrical power', identifying dangers inherent in the operation of nuclear reactors arising from the possibility of radioactive contamination due to accidents relating to coolant leaks, waste disposal, mishandling or meltdown.[45] Williams, a past Dean of the University of Otago's School of Mines and Metallurgy, had in the 1950s been excited about the prospects for uranium mining in New Zealand and had held a prospecting warrant over a piece of land in Buller Gorge. But in his 1977 submission against the introduction of nuclear power to New Zealand, he referred to nuclear power as 'a partially developed technology that is not acceptably safe'.[46]

The Soil Association of New Zealand, the General Practitioners' Society, and the Geological Society of New Zealand were among other scientists and groups of scientists who made submissions against the introduction of nuclear power to New Zealand, with all groups mentioning public safety, or the hazards inherent in the operation of nuclear power stations, as a factor in their opposition. The Institute of Nuclear Sciences' Neil Whitehead, however, made a submission speaking positively about nuclear power, in which he argued that under present safety standards, health risks from nuclear power stations were extremely low and public fears were 'unjustified'.[47]

At the same time as the commission was calling for submissions, the public made clear their views on nuclear power in other ways. In November 1976, Campaign Half Million presented to Parliament a petition with 333,088 signatures, calling for an entirely non-nuclear future for New Zealand: they opposed nuclear power as well as nuclear weapons. Edward Teller, the American scientist known as 'father of the H bomb' ('Try not to call me the father of anything,' said Teller to his Wellington audience. 'My son resents the situation where the H bomb is considered to be his kid brother'), visited New Zealand in 1977 for a week-long study and lecture tour sponsored by the United States Information Service, a propaganda organisation formed by Eisenhower in 1953.[48] Teller visited the Wairakei

This 1976 cartoon, which appeared in the *Sunday News*, shows a small frightened man, representing the New Zealand public, about to be shot from a slingshot wielded by nuclear power proponents onto the point of a large dart protruding from a nuclear power station. Walker is referring to the strong-arm tactics of those supporting the introduction of nuclear power, and the lack of power of those protesting against its introduction. Cartoon by Malcolm Walker, *Sunday News*, 25 Apr. 1976, p. 19.

power station, had morning tea with the New Zealand Atomic Energy Committee, and enjoyed a sherry reception with members of the Christchurch branch of the Royal Society of New Zealand. While hundreds of scientists and government officials attended receptions for the acclaimed scientist, other New Zealanders picketed his talks, protesting his role in the development of nuclear weapons and his advocacy of nuclear power. But even Teller — who did recommend that New Zealand should purchase a research reactor — was cautious about New Zealand's need for nuclear power, saying that the country had to have immediate need for it, and might be better placed to develop its geothermal, wind and wave energy resources.

The report of the Fact Finding Group on Nuclear Power, a scientific investigation that was separate from the Royal Commission, was presented to Government on 31 March 1977, then released to the public. The group's mandate was to report on the possible environmental consequences of nuclear power generation in New Zealand. In contrast to concerns voiced by environmental groups, they reported that under strict supervision and normal operations a nuclear power plant could be 'among the least environmentally

Edward Teller, known as the 'father of the H-bomb', visited New Zealand in 1977. 'Try not to call me the father of anything,' the *New Zealand Listener* reported Teller advising his Wellington audience. 'My son resents the situation where the H bomb is considered to be his kid brother.' *The New Zealand Herald* historic archive H-280212. Courtesy APN New Zealand.

objectionable of the alternatives'.[49] This report presaged many twenty-first-century environmental arguments in favour of nuclear power, not only by highlighting the impact on the landscape of hydro developments and the air pollution problems of thermal power generation, but by mention of the potential of carbon dioxide emissions from fossil-fuelled stations to have 'long-term global effects' that were 'possibly drastic'.[50]

While the royal commission members were hearing submissions, conducting their own research and preparing their report, electricity demand projections were greatly reduced. In the 1977 Report of the Planning Committee on Electric Power Development, updated demand forecasts meant that a nuclear plant was dropped from the fifteen-year plan. By this time, increased power use was no longer seen as a marker of a growing economy — the Government had begun to encourage energy conservation — and demand had not grown as fast as had been predicted. The initial purpose of the commission of inquiry was now, in one sense, obsolete. It was clear that New Zealand did not need to make an immediate decision on whether or not to adopt nuclear power. The commission of inquiry nonetheless completed its investigation and published its report, in the hope that the technical detail amassed and the information about the public debate would be useful when the nuclear option was considered in the future.

The Royal Commission of Inquiry report, *Nuclear Power Generation in New Zealand,* was made public in May 1978. In line with the updated NZED electricity demand projections, the report concluded that 'nuclear power is not justified for New Zealand until about the turn of the century, or even perhaps later'. While this recommendation was the result of reduced demand projections, the commission noted public opposition to the plans for a nuclear power station, describing the history of nuclear power as one of 'official enthusiasm, early public acceptance or apathy, and then of rising opposition'. A survey of New Zealanders at the time showed that only 24–25 per cent favoured nuclear power. The report described nuclear power's early association with military objectives — a connection that had been revealed only in the 1970s — as being responsible for the 'cloud of suspicion and distrust' that enveloped the technology in the minds of the public. They also commented on New Zealanders' general ignorance about nuclear power, something that was still seen as 'an emerging technology, novel and esoteric'. In commenting on the opposition to nuclear power evident in the submissions, the commissioners noted that New Zealand had a strong environmental lobby, but no strong pro-nuclear lobby. The commissioners visited many overseas nuclear installations as part of their inquiry, and were impressed by the officials and engineers they met who were, they said, 'almost unanimous that nuclear power was a necessary and irreplaceable source of the future energy for mankind'.[51]

The report did not reject nuclear power outright, however, suggesting that the question of nuclear power should again be considered in depth by about 1985. It also advised that New Zealand 'maintains and updates its knowledge of nuclear generation as well as evaluating and proving alternative means, so that it is to that extent qualified to avail itself of the nuclear option should it prove desirable'.[52]

Based on the demand forecasts, the inquiry recommended that nuclear power be retained as an option for the future, once indigenous resources had been fully exploited, with an expected commissioning date of 2005–07. Interesting in today's climate is the inquiry's observation that opposition to the environmental impacts of enlarged hydro, geothermal and coal production could, in the future, lead the call for the adoption of nuclear power. 'If New Zealand wants more electricity, and we are sure it will, some environmental impacts will have to be paid,' they said.[53] The commission was far-sighted in recommending such energy-saving measures as encouraging the use of heat pumps and ensuring that all houses have some form of ceiling insulation. These recommendations have only recently been introduced in programmes initiated by the Green Party.

NO NUCLEAR POWER FOR NEW ZEALAND

In the end, nuclear power was rejected primarily for economic reasons, which by now was part of a pattern. In 1956, the UKAEA had cancelled an agreement with New Zealand to produce heavy water at Wairakei because the latest price projections made it economically unfeasible. In 1979, the German company Uranerz would surrender its prospecting licence over uranium resources on the South Island's West Coast after disappointing survey results. New Zealand was now one of the few Western countries not to have some dependence on nuclear technology — whether as an exporter of raw materials for the nuclear industry or as a user of nuclear power. Combined with the ongoing testing of nuclear bombs in the nearby Pacific — which next to no one supported — the stage was set for the widespread opposition to all things nuclear.

New Zealand's rejection of nuclear power felt like a victory for anti-nuclear campaigners. And while, by 1978, there was no immediate need for a nuclear power station, the commission of inquiry's very existence helped to focus attention on the health, safety and environmental issues associated with nuclear technology, and to consolidate public opinion on the matter. Many of the same individuals and groups who opposed the plans for nuclear power would become instrumental in the election of the 1984 Labour Government and the beginning of a 'nuclear-free' New Zealand identity that soon came to embrace nuclear power as well as nuclear weapons.

8

*

A new
national identity
Becoming
'nuclear-free'

*Our isolation and our small size enable us to make a bold and
imaginative initiative that would capture the imagination of the world.*
— RICHARD PREBBLE, 1984[1]

*New Zealand is a nuclear-free country. We reject any strategy for
our defence which relies on nuclear weapons. New Zealand will
not in any way take part in the nuclear arms race or join in any
confrontation between nuclear forces. New Zealand will take no
action which suggests that its security depends on nuclear weapons.*
— PRIME MINISTER DAVID LANGE, 1986[2]

In July 1985, the Greenpeace flagship *Rainbow Warrior* was docked at
its berth in Auckland harbour where it was preparing for a journey to
Moruroa to protest against French nuclear testing. Late on the evening of
10 July 1985, with several crew members sleeping in their cabins or talking
in the messroom, two explosions hit the *Rainbow Warrior*. The ship's
lower regions were flooded, partially submerging the boat and drowning
Portuguese photographer Fernando Pereira, who was trying to retrieve
his camera.

'In no way was France involved,' a representative of the French embassy told the press. But when Police investigations implicated the French Government in this act of terrorism, New Zealanders were outraged. Greenpeace membership grew and anti-nuclear sentiment was cemented as New Zealanders, with their sightings of suspicious activity involving camper vans, Zodiac inflatable dinghies and 'Swiss tourists', took delight in helping the Police to build a case against the suspected French agents.

FRENCH TESTING IN THE PACIFIC

President Charles de Gaulle had chosen 1963 — the same year in which the Partial Test Ban Treaty was signed by the United States, the United Kingdom and the Soviet Union — to announce France's plan to move the French atmospheric testing programme from Algeria's Sahara Desert to the atolls of the South Pacific. New Zealanders marched in protest at the French plans and signed a Campaign for Nuclear Disarmament (CND) petition calling for a complete halt to all nuclear testing in the southern hemisphere, a move initiated by the Australian Labor Party. With the slogan 'No bombs south of the line', the 1963 petition called for a southern hemisphere nuclear-free zone, and, with more than 80,000 signatories, was New Zealand's biggest petition since the women's franchise of 1893. The following year, 1964, China became the world's fifth nuclear power, to condemnation by New Zealand's Prime Minister Keith Holyoake who described the test as 'violating world opinion and greatly increasing the risks of dissemination of nuclear weapons'.[3]

While New Zealand did not have a broad nuclear-free policy — a nuclear power station was on the power plan, and the Government was helping to fund uranium prospecting on the West Coast — it was by now opposed to the continued testing of nuclear weapons. In contrast to the logistical support offered to the British nuclear tests in the Pacific, and, to a lesser extent, support for the American tests, New Zealand was very clearly and vocally opposed to French nuclear testing in the Pacific. Holyoake's Government communicated its opposition to the French plan to test nuclear weapons in the Pacific, and to the continuation of nuclear testing in general, directly to the French Government on several occasions. This opposition was expressed as being on the basis of public concerns about potential dangers to health from radioactive fallout, and because of New Zealand's goal of a comprehensive nuclear test ban.

Despite the protests, France tested its first bomb at French Polynesia's Moruroa Atoll on 2 July 1966. In response to concerns about an increase in radioactive fallout in the region, the National Radiation Laboratory intensified its radiation monitoring programme in the South Pacific. Levels of radioactive fallout were measured in New Zealand and at a chain of Pacific Island stations — in Fiji, Rarotonga, Samoa, Tonga, Niue, and what were then known as the Gilbert and Ellice Islands (now the Republic of Kiribati and Tuvalu), and at Raoul and Penrhyn Islands. Portable gamma-ray monitors were read several times a day at island stations near the test zone. At other stations, daily air-filter samples and weekly rainwater samples were measured for the total fallout content, and milk samples analysed for iodine-131.

In their annual reports on fallout resulting from the French nuclear weapons tests, the National Radiation Laboratory commented that 'there are no internationally accepted levels for the exposure of people to fallout from nuclear weapons testing', but compared levels of radiation attributable to fallout with annual dose limits set by the International Commission on Radiological Protection and with radiation exposure from background radiation.[4] Based on the data collected, and comparison with these levels, the National Radiation Laboratory regularly reported that the French tests 'constituted no public health hazard'.[5] Although the laboratory was dealing with factual data, comparing one figure against another, their refusal to say that the French tests were contributing dangerous levels of radioactive contamination to the Pacific Islands gave some people the impression they were condoning the French testing or even hiding information from the public. In 1972, media attention was given to a Fijian biologist's criticism of the National Radiation Laboratory's interpretation of the monitoring results, along with his comments that radioactive elements like strontium-90 were concentrated in the food chain; for example, as small fish ate contaminated plankton and larger fish ate smaller fish. He advised that a more meaningful monitoring programme would measure levels of radioactive isotopes in the fish that were part of the staple diet of most Pacific Islands people.

New Zealand made diplomatic protests after each of France's tests, and continued to work internationally towards disarmament. In 1968, New Zealand signed the Treaty for the Non-Proliferation of Nuclear Weapons, which was designed to limit the spread of nuclear weapons to other states.

FRENCH ATMOSPHERIC NUCLEAR TESTS IN THE PACIFIC, 1966–74

YEAR	DATE	LOCATION	ESTIMATED YIELD
1966	2 July	Moruroa	20-200 kt
	19 July	Fangataufa	20-200 kt
	11 September	Moruroa	20-200 kt
	24 September	Fangataufa	20-200 kt
	4 October	Moruroa	200 kt-1 Mt
1967	5 June	Moruroa	<20 kt
	27 June	Moruroa	20-200 kt
	2 July	Moruroa	20-200 kt
1968	7 July	Moruroa	20-200 kt
	15 July	Moruroa	200 kt-1 Mt
	3 August	Moruroa	20-200 kt
	24 August	Fangataufa	>1 Mt
	8 September	Moruroa	>1 Mt
1970	15 May	Moruroa	<20 kt
	22 May	Moruroa	200 kt-1 Mt
	30 May	Fangataufa	200 kt-1 Mt
	24 June	Moruroa	<20 kt
	3 July	Moruroa	200 kt-1 Mt
	27 July	Moruroa	<20 kt
	2 August	Fangataufa	20-200 kt
	6 August	Moruroa	200 kt-1 Mt
1971	5 June	Moruroa	20-200 kt
	12 June	Moruroa	200 kt-1 Mt
	4 July	Moruroa	<20 kt
	8 August	Moruroa	<20 kt
	14 August	Moruroa	200 kt-1 Mt
1972	25 June	Moruroa	<20 kt
	30 June	Moruroa	<20 kt
	27 July	Moruroa	<20 kt
1973	21 July	Moruroa	<20 kt
	28 July	Moruroa	<20 kt
	18 August	Moruroa	<20 kt
	24 August	Moruroa	<20 kt
	28 August	Moruroa-aircraft	<20 kt
1974	16 June	Moruroa	<20 kt
	7 July	Moruroa	200 kt-1 Mt
	17 July	Moruroa	<20 kt
	25 July	Moruroa-aircraft	<20 kt
	15 August	Moruroa	20-200 kt
	24 August	Moruroa	<20 kt
	14 September	Moruroa	200 kt-1 Mt

Note: kt = kiloton and Mt = megaton: units of explosive force equivalent to 1000 and 1 million tons of TNT, respectively.

Source: Adapted from *New Zealand at the International Court of Justice: French Nuclear Testing in the Pacific*, New Zealand Ministry of Foreign Affairs and Trade, Wellington, 1996, pp. 48–9.

The United States, the United Kingdom and the Soviet Union signed the treaty in 1968, but its effectiveness was limited by the refusal of China and France — the world's other nuclear powers — to sign.

New Zealand took its opposition to French nuclear testing to the international community. At the United Nations Conference on the Human Environment at Stockholm in June 1972, Duncan McIntyre, the National Government's Minister of Maori and Island Affairs, secured eight co-sponsors for a statement calling for a halt to nuclear weapons tests that could contaminate the environment. Following the tabling of this statement, a draft resolution sponsored by New Zealand and Peru was adopted. It resolved:

> 1. To condemn nuclear weapons tests, especially those carried out in the atmosphere.
> 2. To call upon those states intending to carry out nuclear weapons tests to abandon their plans to carry out such tests as they may lead to further contamination of the environment.[6]

In October, New Zealand's Permanent Representative to the United Nations moved in the First Committee a resolution, with thirteen co-sponsors, calling on all nuclear weapons states to suspend nuclear weapons tests in all environments. The resolution was passed by 106 votes to four, with eight abstentions.

While the Vietnam War dominated the efforts of many in the peace movement, the CND and other groups continued to campaign against French nuclear testing in the Pacific, with street marches in New Zealand and direct protests by boats sailing into the French test zone. In 1972 a 12-metre kauri ketch, the *Vega* — renamed *Greenpeace III* for the trip — sailed from Auckland to the French test area, with support from Greenpeace, the CND, and many Auckland people. In June, as the French tests began, the *Greenpeace III* was rammed by a French Navy minesweeper, damaging the boat, which was towed into Moruroa for repairs. The news did not reach New Zealand for a week, after which it made front-page headlines. Under the headline 'Blast off! Crew of Greenpeace held as bomb tests begin', *The Dominion* reported that *Greenpeace III* had been seized as it entered the French test zone, and its crew were being held near Papeete.[7] As the damaged boat returned to New Zealand for more repairs, other boats sailed towards Moruroa to protest, and the Labour Party announced that if it were in power it would send a frigate to

Moruroa, with Labour Leader Norman Kirk stating in Parliament 'we stand four-square on our policy of opposing nuclear tests, opposing the proliferation of nuclear weapons, bringing about nuclear disarmament'.[8] Meanwhile, the Federation of Labour imposed a ban on French shipping and aircraft, and watersiders refused to load or unload anything to or from France. Post Office workers in Wellington and Auckland placed a temporary ban on telex and cable traffic to and from France and its territories.

This was an election year, and the opposition Labour Party's policy of taking a stronger stance against French nuclear testing may have helped Kirk be elected prime minister of a new Labour Government in November 1972. Once in power, the Kirk Government asked France to postpone any further tests to allow time for discussions. In response, France invited New Zealand to send a scientist to Moruroa. George Roth, retired director of the National Radiation Laboratory, visited the French nuclear test sites from 23 February to 5 March 1973. Roth reported that more atmospheric tests were likely at Moruroa, with underground testing also a possibility, but he also relayed assurances given to him that any official New Zealand protest vessels would come to no harm if they were in the test area.

New Zealand now sought new avenues to oppose the French tests. Growing public concern about the health and environmental effects of nuclear weapons testing led New Zealand and Australia, in 1973, to ask the International Court of Justice in The Hague to challenge the legality of France's atmospheric tests at Moruroa. Before the court had made a ruling on the legality of French atmospheric testing, however, France announced that it would cease atmospheric testing and only test underground in the future. The court decided, by nine votes to six, that since France had stopped atmospheric testing, there was no longer a case to answer.

With this avenue for protest closed, Kirk made good his election-year promise to send a frigate to Moruroa. In 1957, the New Zealand Navy frigates HMNZSS *Pukaki* and *Rotoiti* had sailed to Christmas Island to act as weather ships in support of the British hydrogen bomb tests. In stark contrast, New Zealand Navy frigates were now going to protest against Pacific nuclear testing. In June 1973, Kirk sent the Navy frigate HMNZS *Otago* to Moruroa, to join a small group of protest yachts that had already left from New Zealand, to 'ensure that the eyes of the world are riveted [on] Moruroa'.[9] Kirk, as well as Minister of Defence Arthur Faulkner, and other dignitaries, farewelled the frigate from Auckland. As well as the 243-strong

Navy crew, the frigate carried Immigration Minister Fraser Coleman and a group of journalists, and the National Radiation Laboratory's Jim McCahon, who was on board as radiation safety officer to monitor radiation levels in the air and on the people on board. Faulkner said the frigate — which *The Evening Post* dubbed the 'ban-the-bomb frigate' — would ignore any French attempts to make her leave the testing area: 'she'll proceed on her merry way, exercising her right to peacefully sail the international high seas'.[10]

The HMNZS *Otago* arrived in the test zone in early July, once personal radiation monitors had been distributed to all people on board. On 21 July, a French nuclear bomb was detonated at Moruroa. The *Otago* crew listened to the French countdown, then were on deck a few minutes later, from where they saw the mushroom cloud, which McCahon described as 'a tall spindly stem with a flattened blob on top, a reddish-brown colour against the surrounding white clouds'.[11] Following the explosion, McCahon increased the frequency of radiation measurements but was unable to detect any radioactivity from the small explosion. After the test, the HMNZS *Canterbury* relieved the *Otago* in the test zone, and on 28 July, by which time McCahon had transferred to the *Canterbury*, a second French bomb was detonated. This time, because of the wind direction, the *Canterbury* did detect a small amount of radioactive fallout, equivalent to each person on board receiving a dose of less than 2 millirads, comparable to the radiation received from a diagnostic x-ray. *Canterbury* returned to Auckland on 12 August. The New Zealand frigates had achieved their goal of staging a protest to the French, and of focusing world media attention on French nuclear tests in the Pacific: twice-daily reports by the on-board journalists were published in New Zealand and disseminated around the world.

There were no more atmospheric nuclear bomb tests in the Pacific. In September 1974 France announced that they had completed their programme of atmospheric tests and would now be testing their nuclear weapons underground. This move to a 'safer' form of nuclear testing made them easier for New Zealand scientists to detect. Once the underground testing programme began in 1975, New Zealand scientists found that they could record the explosions on the Pacific seismograph network. 'It was back in 1975 that we saw something very strange on the seismogram recording

Opposite: 'We don't pee in the Atlantic, so don't shit up our Pacific'. Marchers in Wellington city in 1972 protest against French nuclear testing in the Pacific.
1/4-020364-F, *Evening Post* Collection, Alexander Turnbull Library, Wellington, New Zealand.

Jim McCahon, second from the left in the back row, travelled on the HMNZS *Otago*, and then the HMNZS *Canterbury*, as radiation safety officer. An employee of the National Radiation Laboratory since the 1950s, McCahon was pleased to be able to 'add my little bit of protest to the whole thing', but in his diary of his journey he said that amongst the naval officers on board he often felt he was the only person actually protesting against the French nuclear tests, rather than simply following Government orders. Next to McCahon, in the dark shirt, is Immigration Minister Fraser Coleman. The other men, all civilians, are a medical officer and the three journalists. Courtesy Jim McCahon.

in Rarotonga,' Warwick Smith, the DSIR's chief seismologist, later recalled. '"What on earth was that?" we thought — because it didn't look like an earthquake! We subsequently realised that what we had seen was a recording of the first French underground nuclear test in the Pacific.' Underground tests in French Polynesia set up a signal like a sound wave in the ocean that propagated extremely well to the seismogram station in Rarotonga. 'We realised we had quite a sensitive detector of the French nuclear tests.' Smith would announce the test to the Prime Minister's Office, which would contact other countries' top officials and then release the information to the media. 'It was all cloak and dagger stuff for a while', says Smith. 'Then, in the final stages of testing, the French used to make announcements that in less than an hour they would be doing another test.'[12]

FRENCH UNDERGROUND NUCLEAR TESTS IN THE PACIFIC, 1975–96

YEAR	DATE	LOCATION	ESTIMATED YIELD
1975	5 June	Fangataufa	<20 kt
	26 November	Fangataufa	20–200 kt
1976	3 April	Moruroa	<20 kt
	11 July	Moruroa	20–200 kt
	30 October	Moruroa	<20 kt
	5 December	Moruroa	<20 kt
1977	19 February	Moruroa	20–200 kt
	19 March	Moruroa	200 kt–1 Mt
	2 April	Moruroa	<20 kt
	6 July	Moruroa	20–200 kt
	12 November	Moruroa	<20 kt
	24 November	Moruroa	200 kt–1 Mt
	17 December	Moruroa	<20 kt
1978	27 February	Moruroa	<20 kt
	22 March	Moruroa	<20 kt
	25 March	Moruroa	<20 kt
	1 July	Moruroa	<20 kt
	19 July	Moruroa	20–200 kt
	26 July	Moruroa	<20 kt
	2 November	Moruroa	<20 kt
	30 November	Moruroa	200 kt–1 Mt
	17 December	Moruroa	<20 kt
	19 December	Moruroa	20–200 kt
1979	1 March	Moruroa	20–200 kt
	9 March	Moruroa	20–200 kt
	24 March	Moruroa	20–200 kt
	4 April	Moruroa	20–200 kt
	18 June	Moruroa	20–200 kt
	29 June	Moruroa	200 kt–1 Mt
	25 July	Moruroa	200 kt–1 Mt
	28 July	Moruroa	<20 kt
	19 November	Moruroa	<20 kt
	22 November	Moruroa	<20 kt
1980	23 February	Moruroa	<20 kt
	3 March	Moruroa	<20 kt
	23 March	Moruroa	200 kt–1 Mt
	1 April	Moruroa	20–200 kt
	4 April	Moruroa	20–200 kt
	16 June	Moruroa	200 kt–1 Mt
	21 June	Moruroa	20–200 kt
	6 July	Moruroa	20–200 kt
	19 July	Moruroa	200 kt–1 Mt
	25 November	Moruroa	<20 kt
	3 December	Moruroa	200 kt–1 Mt
1981	27 February	Moruroa	<20 kt
	6 March	Moruroa	<20 kt
	28 March	Moruroa	20–200 kt

221

YEAR	DATE	LOCATION	ESTIMATED YIELD
1981 (cont.)	10 April	Moruroa	20–200 kt
	8 July	Moruroa	20–200 kt
	11 July	Moruroa	<20 kt
	18 July	Moruroa	<20 kt
	3 August	Moruroa	200 kt-1 Mt
	6 November	Moruroa	<20 kt
	11 November	Moruroa	20–200 kt
	5 December	Moruroa	20–200 kt
	8 December	Moruroa	20–200 kt
1982	20 February	Moruroa	<20 kt
	24 February	Moruroa	<20 kt
	20 March	Moruroa	20–200 kt
	23 March	Moruroa	<20 kt
	27 June	Moruroa	<20 kt
	1 July	Moruroa	200 kt-1 Mt
	21 July	Moruroa	<20 kt
	25 July	Moruroa	200 kt-1 Mt
	27 November	Moruroa	<20 kt
1983	19 April	Moruroa	200 kt-1 Mt
	25 April	Moruroa	<20 kt
	25 May	Moruroa	200 kt-1 Mt
	18 June	Moruroa	<20 kt
	28 June	Moruroa	20–200 kt
	20 July	Moruroa	20–200 kt
	4 August	Moruroa	200 kt-1 Mt
	3 December	Moruroa	<20 kt
	7 December	Moruroa	20–200 kt
1984	8 May	Moruroa	<20 kt
	12 May	Moruroa	200 kt-1 Mt
	12 June	Moruroa	20–200 kt
	16 June	Moruroa	200 kt-1 Mt
	27 October	Moruroa	20–200 kt
	2 November	Moruroa	200 kt-1 Mt
	1 December	Moruroa	<20 kt
	6 December	Moruroa	200 kt-1 Mt
1985	30 April	Moruroa	20–200 kt
	8 May	Moruroa	200 kt-1 Mt
	3 June	Moruroa	20–200 kt
	7 June	Moruroa	20–200 kt
	24 October	Moruroa	<20 kt
	26 October	Moruroa	200 kt-1 Mt
	24 November	Moruroa	20–200 kt
	26 November	Moruroa	200 kt-1 Mt
1986	26 April	Moruroa	20–200 kt
	6 May	Moruroa	<20 kt
	27 May	Moruroa	20–200 kt
	30 May	Moruroa	200 kt-1 Mt
	10 November	Moruroa	<20 kt

YEAR	DATE	LOCATION	ESTIMATED YIELD
1986 (cont.)	12 November	Moruroa	20–200 kt
	6 December	Moruroa	<20 kt
	10 December	Moruroa	200 kt–1 Mt
1987	5 May	Moruroa	20–200 kt
	20 May	Moruroa	200 kt–1 Mt
	6 June	Moruroa	20–200 kt
	21 June	Moruroa	200 kt–1 Mt
	23 October	Moruroa	200 kt–1 Mt
	5 November	Moruroa	20–200 kt
1988	11 May	Moruroa	200 kt–1 Mt
	25 May	Moruroa	200 kt–1 Mt
	16 June	Moruroa	<20 kt
	23 June	Moruroa	20–200 kt
	25 October	Moruroa	<20 kt
	5 November	Moruroa	200 kt–1 Mt
	23 November	Moruroa	200 kt–1 Mt
	30 November	Fangataufa	200 kt–1 Mt
1989	11 May	Moruroa	20–200 kt
	20 May	Moruroa	<20 kt
	3 June	Moruroa	200 kt–1 Mt
	10 June	Moruroa	200 kt–1 Mt
	24 October	Moruroa	200 kt–1 Mt
	31 October	Moruroa	20–200 kt
	20 November	Moruroa	20–200 kt
	27 November	Moruroa	200 kt–1 Mt
1990	2 June	Moruroa	20–200 kt
	7 June	Moruroa	20–200 kt
	26 June	Fangataufa	200 kt–1 Mt
	4 July	Moruroa	20–200 kt
	14 November	Fangataufa	200 kt–1 Mt
	22 November	Moruroa	200 kt–1 Mt
1991	7 May	Moruroa	<20 kt
	18 May	Moruroa	200 kt–1 Mt
	29 May	Fangataufa	200 kt–1 Mt
	16 June	Moruroa	200 kt–1 Mt
	5 July	Moruroa	<20 kt
	15 July	Moruroa	200 kt–1 Mt
1995	5 September	Moruroa	~20 kt
	1 October	Fangataufa	~150 kt
	27 October	Moruroa	~60 kt
	21 November	Moruroa	~60 kt
	27 December	Moruroa	~30 kt
1996	27 January	Fangataufa	<120 kt

Note: kt = kiloton and Mt = megaton: units of explosive force equivalent to 1000 and 1 million tons of TNT, respectively.

Source: Adapted from *New Zealand at the International Court of Justice: French Nuclear Testing in the Pacific*, New Zealand Ministry of Foreign Affairs and Trade, Wellington, 1996, pp. 48–9, with 1995–96 data from www.ratical.org/ratville/nukes/testChrono95-8.html and http://nuclearweaponarchive.org.

Even after France took its nuclear testing programme underground — thus avoiding the release of radioactive isotopes like strontium-90 and caesium-137 directly into the atmosphere — there remained a general public suspicion that the tests were harmful. New Zealand, along with Australia and various small Pacific countries, continued to protest to France after each test, and New Zealanders protested on street marches and boycotted French goods. The National Radiation Laboratory continued to monitor the impact of the nuclear explosions, including tests designed to detect any venting to the atmosphere from the underground nuclear tests, and issued annual reports on environmental radioactivity in New Zealand and the Pacific. Partly because of continuing challenges by Greenpeace and other environmental voices, the National Radiation Laboratory was no longer the universally trusted voice that it was when it first started issuing environmental monitoring reports in the 1960s. Greenpeace pointed out in 1990 that by 1977 the National Radiation Laboratory had been monitoring the atmosphere since 1957 and had issued 50 reports in which they regularly declared the Pacific nuclear tests did not constitute a public health hazard.

As well as making specific protests to France after each nuclear test, the Labour Government also worked towards wider disarmament issues, pushing for a South Pacific nuclear-weapons-free zone and a halt to weapons testing in general. On 31 October 1975, the New Zealand Ambassador Malcolm Templeton introduced to the United Nations a resolution, co-sponsored with Fiji and Papua New Guinea, calling for a South Pacific nuclear-weapons-free zone. In November 1975, however, Labour lost the election to National. Robert Muldoon became prime minister, and the initiative was abandoned.

Then, in February 1976, Muldoon declared that the National Party caucus was unanimous in agreeing with his policy of lifting the ban on visits to New Zealand by nuclear warships: the first visit could be that very year.

VISITS FROM NUCLEAR-POWERED SHIPS

American nuclear-powered vessels — a submarine, two cruisers and an aircraft carrier — had visited New Zealand in 1960 and 1964, but from 1965 to 1975 New Zealand had made it clear to the United States authorities that they were not welcome. The issue at this stage was not public opposition to nuclear ship visits (although there was growing opposition from environmental and peace groups), but rather the issue of liability in the case of an accident.

Kevin Clements has described Labour supporters in the mid-1970s as being 'fervently anti-nuclear for principled and political reasons', whereas the new National Government cast itself as 'politically realistic in foreign and defence policies and not subject to the "woolly minded" schemes of Labour in promoting unrealistic concepts such as a Pacific-wide nuclear-weapons free zone'.[13] In 1976, therefore, National Prime Minister Robert Muldoon advised that nuclear-powered ships could once again visit New Zealand so long as the owning nation accepted full responsibility for the ship, and the ship complied with the safety standards of the port it was visiting, by now updated in the new Code of Practice for Nuclear Powered Shipping prepared by the New Zealand Atomic Energy Committee. Muldoon saw any continuation of Labour's ban on nuclear ship visits as 'nonsensical and a danger to the continuation of the ANZUS alliance'.[14]

NUCLEAR-POWERED SHIP VISITS TO NEW ZEALAND

DATES	VESSEL	TYPE	PORT
19–22 April 1960	USS *Halibut*	Submarine	Auckland
24–27 April 1960	USS *Halibut*	Submarine	Wellington
8–9 September 1964	USS *Long Beach*	Cruiser	Wellington
8–9 September 1964	USS *Bainbridge*	Cruiser	Wellington
8–9 September 1964	USS *Enterprise*	Aircraft carrier	Wellington
27 August–2 September 1976	USS *Truxtun*	Cruiser	Wellington
1–5 October 1976	USS *Long Beach*	Cruiser	Auckland
16–22 January 1978	USS *Pintado*	Submarine	Auckland
19–24 January 1979	USS *Haddo*	Submarine	Auckland
22–29 September 1980	USS *Truxtun*	Cruiser	Wellington
25–29 May 1982	USS *Truxtun*	Cruiser	Wellington
3–8 August 1983	USS *Texas*	Cruiser	Auckland
10–15 August 1983	USS *Texas*	Cruiser	Wellington
9–14 November 1983	USS *Phoenix*	Submarine	Auckland
23–30 March 1984	USS *Queenfish*	Submarine	Auckland

Source: Adapted from Special Committee on Nuclear Propulsion, *The Safety of Nuclear Powered Ships*, Department of the Prime Minister and Cabinet, Wellington, 1992, p. 183.

April 10th, 1968. 51 die in Wahine Disaster.

Had this ship been nuclear powered, thousands could have died.

Campaign Against Nuclear Warships poster, 1976. GH014460, Te Papa, Wellington, New Zealand.

In response to Muldoon's plans to resume nuclear ship visits, Labour MP Richard Prebble sought recognition of a South Pacific nuclear-free zone with the August 1976 introduction of his Nuclear Free Zone (New Zealand) Bill. Three weeks later, another Labour MP, Edward Isbey, introduced the United Nations Nuclear Free Zone Resolution Adoption Bill, which would adopt the United Nations resolution on a nuclear-free zone in the South Pacific that had been passed in December 1975. Both Bills were rejected by the Government and not allowed a first reading.

The Muldoon Government's invitation for the nuclear-powered cruiser USS *Truxtun* to visit Wellington was a further rejection of the proposed South Pacific nuclear-weapons-free zone. On 27 August 1976, as the *Truxtun* entered Wellington harbour, she was met by a small Peace Squadron — part of a group formed to co-ordinate and inspire boat owners to use their craft to prevent nuclear ships from entering New Zealand ports — whose boats were outnumbered by security boats. In protest at the *Truxton*'s visit, members of the Harbour Employees Union and the Watersiders' Union stopped work soon after the *Truxtun* entered the harbour. Ferry sailings were cancelled (even though it was school holidays), and the *Truxtun* was forced to anchor in the stream for her six-day visit. Muldoon and a group of Cabinet ministers and MPs were not bothered by the protest. They lunched with the captain on board the ship, after which Muldoon made fun of the protesters' objections by telling media 'I didn't see anyone with four thumbs'.[15]

By the time of the next nuclear ship visit, the Auckland Peace Squadron had 50 craft registered, crewed and skippered by what Tom Newnham has called 'members of Parliament, company directors and executives, radical activists and many ordinary members of the public who had never participated in protest before'.[16] As the USS *Long Beach* approached Auckland early on the morning of 1 October 1976, she was met by members of the Peace Squadron, forcing her to slow down and stop while the protest boats were dealt with by the cruiser's Navy and Police escort. The *Long Beach* eventually made its way into the harbour, through and past more than 150 protesting small craft — yachts, launches, dinghies and canoes. In Auckland, 3000 people marched down Queen Street in protest at the visit of the nuclear cruiser, and in Parliament politicians spent two hours debating the pros and cons of the visit following a motion by Labour MP Richard Prebble that the anchorage of the *Long Beach* in the harbour between Kings Wharf and the Devonport naval base was unsafe. While the Peace Squadron's plan to stop the *Long Beach* berthing in Auckland failed, this first full-scale demonstration by the anti-nuclear movement attracted strong media attention and set the pattern for later protests.

It was not just protestors who sprang into action when a nuclear ship arrived in New Zealand waters. The Ministry of Civil Defence's 1976 Public Safety Plan for the Port of Wellington required that the day before the scheduled arrival of a nuclear-powered ship, the Minister of Civil Defence, in conjunction with the Police, Defence Department and local authorities, would establish a Public Safety Operational Headquarters in the basement of the Beehive, under the supervision and control of the Regional Commissioner of Civil Defence. As part of being in a state of 'relaxed readiness' for an incident or accident, a tugboat crew would have

The look-out tower of Ian and Clare Athfield's house was painted with an anti-nuclear slogan when the USS *Truxtun* visited Wellington in 1976. *The Dominion* and *Sunday Times* photograph. Courtesy Fairfax NZ Limited.

on board protective clothing, individual dosimeters (to measure radiation exposure), potassium-iodate tablets* and service respirators for their crew. The Department of Health would also issue the Ministry of Defence, the Police, the Ministry of Transport, the Wellington Free Ambulance, the Wellington Regional Fire Board and the Wellington Harbour Board with potassium-iodate tablets. Police, who would also be supplied with protective clothing, individual dosimeters and service respirators, were required to have available roadblock signs saying *STOP — RADIOACTIVE HAZARD AREA*. The Wellington civil defence plan first sprang into action for the 1976 visit of the USS *Truxtun*. In the days before the ship arrived, the National Radiation Laboratory installed radiation monitoring devices at 20 sites around Wellington harbour, and two busloads and a planeful of Police officers arrived from Auckland. The day before the ship arrived, the national civil defence nerve centre was activated under Parliament buildings, with *The Dominion* reporting the arrival of five Land Rovers carrying troops with technical equipment and sleeping bags.

The next nuclear ship visit was in January 1978, when the USS *Pintado* was confronted by about 80 protest vessels — canoes, surfsailers, motorboats and yachts — on its approach to Auckland. This time, anticipating the nuclear-powered submarine's reception, the HMNZS *Waikato* helped the *Pintado* navigate her way through the protest craft into Auckland. Along with the frigate was a fleet of Navy and Police vessels and two Navy helicopters, which escorted the submarine and harassed protest vessels. The protesters had wide support: the next day, *The New Zealand Herald* reported that a flow of letters to the editor on the subject was running at a ratio of nine to seven against nuclear-powered ships coming to New Zealand.

The Labour Party fought the 1978 election on a promise to close the country's ports and airports to all nuclear-powered and nuclear-weapon-carrying craft, but the Muldoon Government was re-elected. The visit of the nuclear submarine USS *Haddo* on 19 January 1979 met with a large protest fleet, including yachts, dinghies, kayaks and surfboards. On land, street protesters — whom *The New Zealand Herald* described as 'predominantly young, generally sandaled and mostly white' — marched through central

---

* The tablets were for taking in the event of a nuclear accident: iodine concentrates in the thyroid gland and taking a non-radioactive form of iodine would protect the individual from the accumulation of cancer-causing radioactive iodine-131 released as a fission product.

Protestor Stephen Sherie on the bow of the USS *Haddo* in Auckland harbour, 20 January 1979.
EP-Navy-Warships-USS Haddo-02, Alexander Turnbull Library, Wellington, New Zealand.

Auckland then massed at the waterfront, chanting 'One, two, three, four, we don't want your nuclear war.' The American commander of the *Haddo* described the approach to Auckland as 'frightening because of the danger the protesters put themselves in'. The head of the Auckland Police district was not so polite, describing them as 'seagoing hoodlums rather than protestors'.[17] Two men managed to climb aboard the submarine, while fellow protesters threw yellow paint bombs at the vessel, earning the headline 'Nuclear Haddo a "yellow submarine" as protestors get far too close' in Wellington's *Evening Post*.[18]

The National Party remained in power and American nuclear cruisers and submarines continued to visit Auckland and Wellington. The USS *Truxtun* visited Wellington in September 1980 and May 1982, the USS *Texas* visited Auckland and Wellington in August 1983, and the USS *Phoenix* visited Auckland in November 1983. Peace Squadron boats on the harbour and protesters marching through the city streets greeted each nuclear ship. The last nuclear ship to visit New Zealand was the submarine USS *Queenfish*,

The USS *Haddo* is met in Auckland harbour by a flotilla of protesting boats.
EP-Navy-Warships-USS Haddo-01, Alexander Turnbull Library, Wellington, New Zealand.

which docked in Auckland in March 1984. The nuclear-powered submarine was confronted by a Peace Squadron protest fleet that *The New Zealand Herald* described as consisting of '58 yachts, 12 canoes, about 30 other small craft, a surf sailor – and a naked man on a surfboard'.[19] The *Auckland Star* described the protest as the 'most spectacular' yet, with boats firing red flares and flying black flags, and protesters chanting and beating drums as the *Queenfish* slid into her mooring.[20]

This nuclear issue was different to those the country had dealt with in the past. Plans for a uranium industry on the West Coast and a nuclear power station on the Kaipara Harbour were initiated by New Zealanders and had clear benefits for the nation. In contrast, many people saw the nuclear ships issue as another country imposing its foreign policy on New Zealand, in a way that put New Zealanders at risk. The revisiting of the nuclear-powered ships debate re-ignited the anti-nuclear movement and gave it a broader-based support. In 1981, the New Zealand Nuclear-Free Zone Committee was established in Christchurch, and began organising a signature campaign and publicity that included bumper stickers and badges. The committee encouraged local groups to declare themselves nuclear-free. As well as broad opposition to the continuing nuclear arm's race, there were genuine concerns about the potential health impact of American nuclear ship visits and the continuing French nuclear-testing programme.

In 1983, France invited a number of scientists from South Pacific countries to visit Moruroa Atoll to study the effects of the underground nuclear-test programme. (New Zealand had been pressing France for some years to allow scientists to visit.) In October 1983, the National Radiation Laboratory's director, Hugh Atkinson, led a scientific mission to Moruroa. The team included five scientists from New Zealand, Australia and Papua New Guinea, including Atkinson and Andrew McEwan, Atkinson's successor as director of the National Radiation Laboratory. During their four days on Moruroa, the scientists collected air, water, shellfish, algae, coral, reef fish, and soil and vegetation samples from on and around the atoll, as well as plankton and ocean fish from outside the reef. The samples were then analysed by either the National Radiation Laboratory in Christchurch or the Australian Atomic Energy Commission laboratories at Lucas Heights.

The scientists' report, released in July 1984, was reassuring. While it concluded that 20 per cent of long-lived fission products contributing radiation doses to New Zealanders came from French nuclear tests, as opposed to the earlier mostly northern hemisphere tests, it said that resulting radiation levels were not harmful. Even in French Polynesia, the report concluded, radiation doses 'are lower than world average levels and do not lead to any expectation that radiation-induced diseases would be detectable'.[21] Newspapers, however, picked up on the report's comments that in a worst-case scenario, fracturing of the volcanic rock beneath the atoll could lead to radioactivity leaking within five years. Some people, however, saw the report's finding that the tests were not causing dangerous levels of radioactivity as fuelling French claims that the tests were harmless. Greenpeace released a counter-paper calling for independent scientists to be allowed access to the data, and arguing that new, more comprehensive, data should be collected on radioactive contamination of the limestone and coral beneath and within the lagoon, and for an anecdotal survey of the health of Polynesian people.

Greenpeace was also beginning to focus media attention on the effect of the American nuclear-testing programme on the Marshall Islanders. There were reports of multiple miscarriages, babies born with birth defects, and high rates of cancer amongst the islanders, and in May 1985 a Greenpeace ship, the *Rainbow Warrior*, evacuated 320 people from the radiation-contaminated Rongelap Atoll to a nearby atoll.

THE FOURTH LABOUR GOVERNMENT AND A 'NUCLEAR-FREE' NEW ZEALAND

David Lange became head of the opposition Labour Party in 1983. He later recalled that he 'took it for granted that we would ban nuclear weapons from New Zealand as soon as Labour was elected'.[22] This ended up being one of the defining issues of the next general election.

In June 1984, Richard Prebble again attempted to legislate a nuclear-free status for New Zealand by introducing a Private Member's Bill. The Nuclear Free New Zealand Bill, Prebble told Parliament, would ban visits by nuclear-armed or nuclear-powered ships, prohibit the building of nuclear reactors or the dumping of nuclear waste in New Zealand, and would completely ban all nuclear weapons. 'Our isolation and our small size,' Prebble said, 'enable us to make a bold and imaginative initiative that would capture the imagination of the world.'[23] The National Government was against the Bill, saying it would spell an end to the ANZUS alliance, the security agreement between Australia, New Zealand and the United States signed in 1951. But when the Bill was put to the vote, National MPs Marilyn Waring and Mike Minogue voted with the Labour Party and Social Credit in favour of a second reading. The National Government, which had a majority of only one in the House, won the vote (there were 40 votes against the Bill and 39 for it) only because two former Labour MPs, now acting as independents, voted with them against the Bill. The next day, National MP Marilyn Waring withdrew from the party caucus, saying she would no longer vote with the National Party on disarmament matters. Although the Nuclear Free New Zealand Bill had been narrowly defeated, Prime Minister Robert Muldoon acknowledged that the Government's majority was now uncertain, and called a snap election to be held in one month's time, on 14 July 1984. According to Lange, who succeeded Muldoon as prime minister, and George Gair, a minister in Muldoon's Cabinet, Muldoon's Government had run out of money and calling an early election avoided the

Naomi and David Lange at Labour Party headquarters in Mangere on election night, 16 July 1984, photographed by *Evening Post* staff photographer Phil Reid. EP/1984/3357/23-F, Alexander Turnbull Library, Wellington, New Zealand.

need to produce a Budget that would reveal this fact. Whether the nuclear-free issue was a reason or an excuse for a snap election is debateable, but it was certainly at least a catalyst.

Nuclear issues received a lot of attention in the month-long election campaign. Of the four political parties that fought the 1984 general election, only the National Party did not promise to ban nuclear-armed and nuclear-powered vessels from New Zealand ports. The Labour Party's stated policy on international affairs was to reaffirm its prohibition of visits by nuclear-armed and nuclear-powered warships, seek the establishment of a South Pacific nuclear-weapons-free zone, and prohibit the dumping of nuclear wastes and the testing of nuclear weapons in the Pacific.

In the general election of 14 July 1984, Labour won a landslide victory and David Lange became prime minister. Labour's nuclear-free policy was a reflection of what by now was the mood of New Zealanders. By November 1984, 94 areas of New Zealand had declared themselves nuclear-free, accounting for 2,075,747 people, or 65 per cent of the population.

The Ministry of Foreign Affairs, like the National Party, was concerned about the impact of Labour's nuclear-free policy on ANZUS. The ministry advised Lange's Government that pursuing its nuclear-free policy would be harmful to New Zealand's relationship with the United States and could have a negative impact on New Zealand's security arrangements and economy. Lange wanted New Zealand to remain in the ANZUS alliance, and he believed he could get around the issue by selecting an American ship to visit New Zealand that would not impinge on New Zealand's nuclear-free policy; a ship that was neither nuclear-powered or nuclear-armed. Lange discussed the issue with United States Secretary of State George Shultz in July and September 1984, then in November he sent Ewan Jamieson, the New Zealand Chief of Defence Staff, to Hawaii to discuss with the United States military authorities a suitable ship. The United States had a strict 'neither confirm nor deny' policy when it came to questions about whether or not its ships were carrying nuclear weapons, but they understood New Zealand's situation and suggested the USS *Buchanan*, an aged oil-fired Navy destroyer. Jamieson returned to New Zealand and told Lange the USS *Buchanan* would be the ideal ship to visit New Zealand. In January 1985, the United States made a request to the New Zealand Government for the USS *Buchanan* to visit New Zealand. The request was leaked to the media and there was wide speculation about the nuclear capabilities of the vessel. By now Lange was holidaying in the Tokelau Islands and was unable to be contacted; public officials and his Government had to consider the request without him.

Lange returned to New Zealand to advice from the Ministry of Foreign Affairs that the request be accepted, and to a report from Deputy Prime Minister Geoffrey Palmer — agreed to by Cabinet — which advised that, because they could not conclusively say whether the USS *Buchanan* was nuclear-armed or not, the request be refused. Lange endorsed Palmer's report. In a private meeting between Lange and United States Ambassador H. Monroe Brown, the United States was advised that the visit of the USS *Buchanan* could not proceed, as the ship did not conform to New Zealand policy. Lange suggested that a different ship — one that was not nuclear-capable — be sent in its place. New Zealand's refusal to give access to the USS *Buchanan* was announced publicly on 4 February to shock and anger from the United States, who had been under the impression that they had an informal agreement with Lange that their request would be accepted. Support from the New Zealand public, however, was strong. The Lange Government's refusal of the *Buchanan* visit coincided with 15,000 people marching down Queen Street in Auckland chanting 'If in doubt, keep it out!'

The United States responded by greatly reducing defence and intelligence co-operation with New Zealand under ANZUS, and by making it clear that, if New Zealand wished to have an effective defence relationship with the United States, it must accept American nuclear weapons. New Zealand was portrayed as 'anti-ANZUS, anti-American, even anti-Western' with some Americans having the attitude that if 'you're not with us in every particular, you must be against us'.[24] Despite both Australia and the United States withdrawing from planned ANZUS military exercises, Lange stated that, although New Zealand did not wish to be defended by nuclear weapons, the country remained committed to ANZUS.

New Zealanders' support for the nuclear-free policy was strengthened by the United States' response to New Zealand's stand. Foreign policy expert Stuart McMillan later claimed that the nuclear ships ban 'assumed elements of an assertion of national identity. In the minds of some people, what it meant to be a New Zealander, partly, was to live in a nuclear free country.'[25] On 18 February 1985, a nationwide public opinion poll conducted for *The Dominion* revealed that 56 per cent of respondents were against nuclear-armed warships visiting New Zealand. In addition, 42 per cent of respondents were against visits from warships that were nuclear-powered.

New Zealand's ban on nuclear ships, and the United States' response to it, put New Zealand's nuclear-free policy into world headlines and attracted the notice of other governments. In March 1985, Lange's televised appearance at the Oxford Union debate at Oxford University also gained international media attention and boosted New Zealand's pride in their prime minister. At the 1 March debate, Lange spoke in the affirmative, against American senator and moral majority leader Jerry Falwell, for the proposition 'that nuclear weapons are morally indefensible'. His response to a young Rhodes Scholar who asked how Lange could justify New Zealand's continued membership of ANZUS — Lange suggested he would answer if the young man would hold his breath for a moment, adding 'I can smell the uranium on it as you lean towards me!' — was received with laughter and applause from the audience.[26] Lange's intellect was praised by the British media, and his side won the debate 298 votes to 250.

THE BOMBING OF THE *RAINBOW WARRIOR*

The bombing of the *Rainbow Warrior* in 1985 further reinforced New Zealand's emerging identity as an independent and nuclear-free nation willing to stand up for its principles in the face of bigger, more powerful nations.

Since the New Zealand Government's 1973 frigate protest, many private yachts had continued to leave from New Zealand to protest against the French nuclear tests at Moruroa. On 10 July 1985, the *Rainbow Warrior* was berthed in Auckland Harbour in preparation for a similar journey. The bombing of the Greenpeace ship captured the attention of the nation and led to the biggest Police operation in New Zealand history. When Police investigations revealed the likely participation of a team of agents from the French intelligence service, the media attention intensified. On 23 July two French agents, later identified as Major Alain Marfart and Captain Dominique Prieur, were charged with arson and the murder of Pereira. Police investigators gathered evidence from New Zealand, France and Interpol. In turn, the French Government conducted its own 'investigation' and initiated a programme of disinformation, in which they suggested Communist involvement in Greenpeace and in the bombing itself.

The *Rainbow Warrior* in Auckland harbour after bombing by French secret service agents. © Greenpeace/ John Miller (www.greenpeace.org/new-zealand/en/about/ships/the-rainbow-warrior).

The French prime minister, however, later acknowledged that agents of the French secret services were responsible for sinking the *Rainbow Warrior*, acting under orders to that effect. On the first day of Mafart and Prieur's trial, 4 November 1985, their charges were reduced from murder and arson to manslaughter and wilful damage. The French agents pleaded guilty and were subsequently each sentenced to ten years' imprisonment.

Historian Michael King described the bombing of the *Rainbow Warrior* as being not an isolated incident, but part of the context of '40 years' use of the Pacific Ocean as a nuclear testing and dumping ground' and the result of France's belief 'that it had the right to use any means to safeguard its own testing programme'.[27] This act of State-sponsored terrorism, followed by France's continuation with the nuclear test series at Moruroa while the French agents were awaiting trial, outraged the New Zealand public and cemented support for the anti-nuclear movement. In the months after the bombing, Greenpeace membership grew and donations to the organisation reached more than $200,000. As Deputy Prime Minister Geoffrey Palmer said in an October speech in The Hague: 'this outrageous incident has only helped to strengthen opposition in New Zealand and elsewhere to nuclear testing'.[28] The challenges to New Zealand's nascent nuclear-free policy served only to entrench it.

France was not contrite: the French Government retaliated to New Zealand's imprisonment of Mafart and Prieur by blocking or delaying imports of a range of New Zealand products. In July 1986, amid claims that France would veto New Zealand butter exports to the European Economic Community if the agents were not released, Mafart and Prieur left New Zealand. Under an agreement negotiated by the United Nations Secretary General, they were flown to Hao Atoll, 450 kilometres north-north-west of Moruroa, where they were to remain for three years. In exchange, France was to pay New Zealand US$7 million compensation and issue a formal letter of apology. In December 1987, however, Mafart was repatriated to Paris for medical treatment, where he was promoted to the rank of major. Prieur returned to France in April 1988 amid reports she was pregnant. The New Zealand Government registered a formal protest. In May 1990, an international arbitration tribunal agreed that France had breached its international legal obligations by removing the agents from Hao, and recommended that France contribute US$2 million to a fund to promote friendly relations between the two countries.

A NUCLEAR-FREE ZONE AT LAST

Other countries joined New Zealand in promoting a nuclear-free South Pacific. On 6 August 1985, one month after the bombing of the *Rainbow Warrior* (and 40 years after an atomic bomb was dropped on Hiroshima), New Zealand was one of eight countries to sign the South Pacific Nuclear Free Zone Treaty at the South Pacific Forum in Rarotonga. Initiated by Australia and New Zealand, the Rarotonga treaty sought to establish the world's third nuclear-weapons-free zone by preventing the storage, dumping, manufacture and testing of nuclear weapons in the South Pacific.

At a Labour Party regional conference the next month, Minister of Defence Frank O'Flynn moved a resolution calling for the 'urgent implementation of [anti-nuclear] legislation'.[29] His resolution was successful, and David Lange, as Minister of Foreign Affairs, introduced the draft legislation on 10 December 1985 as the New Zealand Nuclear Free Zone, Disarmament and Arms Control Bill. It had its first reading and was referred on to the Foreign Affairs and Defence Committee. On introducing the Bill, Lange said it fulfilled New Zealand's obligations under the South Pacific Nuclear Free Zone Treaty and reflected 'the intention of the signatories to the treaty that the destabilising elements of nuclear confrontation not be allowed to intrude into this region'.[30]

Publicly, anti-nuclear sentiments continued to grow. In April 1986, an accident at the Chernobyl nuclear power plant in the Soviet republic of Ukraine killed 31 people, caused the relocation of more than 100,000 people, and sent a plume of radioactivity across Europe. The incident, the world's worst nuclear reactor accident, received international publicity and further solidified public fear and antipathy to all things nuclear. A 1986 Defence Committee of Enquiry ran a national poll which revealed that: 92 per cent of respondents were opposed to nuclear weapons being stationed in New Zealand; 73 per cent wanted a nuclear-free defence policy; and 66 per cent wanted nuclear-armed ships banned from New Zealand ports (41 per cent wanted nuclear-powered ships banned, too). This was in contrast to a similar survey, conducted in the late 1970s, that showed that more than 60 per cent of respondents were in favour of visits by American nuclear-armed warships.

As warned by the Ministry of Foreign Affairs, New Zealand's nuclear-free policy spelt the end of the ANZUS agreement. In June 1986, United States Secretary of State George Shultz told the press that New Zealand and the

In June 1986, the United States made it clear that New Zealand would be kicked out of the ANZUS agreement. Here, David Lange is seen being booted out of Cafe Anzus by George Shultz, while Australian Prime Minister Bob Hawke shares a drink with United States President Ronald Reagan. Cartoon by Tom Scott, 5 Jul. 1986, A-312-4-001, Alexander Turnbull Library, Wellington, New Zealand.

United States 'part company as friends, but we part company', indicating that in order to be treated as an ally by the United States New Zealand would have to accept that from time to time she would be visited by a United States warship that was nuclear-armed.[31] In August, the United States announced it was formally suspending its security commitment to New Zealand under ANZUS pending adequate corrective measures over New Zealand's stance on nuclear ship visits.

The South Pacific Nuclear Free Zone Treaty was ratified by New Zealand on 13 November 1986 and came into force on 11 December 1986. Combined with the nuclear-free zones of Latin America to the east, and Antarctica to the south, it meant that 40 per cent of the Earth's surface was declared nuclear-free.

The New Zealand Nuclear Free Zone, Disarmament and Arms Control Act entered into force on 8 June 1987. The Act's stated purpose was 'to establish in New Zealand a Nuclear Free Zone, to promote and encourage an active and effective contribution by New Zealand to the essential process of disarmament and arms control' and to implement the South Pacific Nuclear Free Zone Treaty and four other international treaties and conventions relating to disarmament and arms control. The Act established a nuclear-free zone comprising all New Zealand land, water and airspace out to the limits of the territorial sea of New Zealand. Within this nuclear-free zone there was a full prohibition on the testing or transporting or stationing of nuclear weapons. Clause 11 of the Act banned the entry into the internal waters of New Zealand of any ship 'whose propulsion is wholly or partly dependent on nuclear power'. The Act also established a Public Advisory Committee on Disarmament and Arms Control to advise the minister of foreign affairs and the prime minister on matters related to disarmament and arms control and the implementation of the Act.

The Act amended the Marine Pollution Act 1974, as well, making it an offence to dump any radioactive waste or other radioactive matter into New Zealand waters. On behalf of the Ministry of Health, the National Radiation Laboratory Director Andrew McEwan had earlier made a submission against this aspect of the Act, pointing out that combining in one Bill legislative controls on nuclear-powered ships and the controlled disposal of radioactive waste 'may reinforce public misconceptions relating to hazards associated with useful applications of radioactive materials and nuclear technology' and re-stated the National Radiation Laboratory's stance that the existing Convention on the Prevention of Marine Pollution by Dumping of Wastes and Other Matters, to which New Zealand was a signatory, was 'a satisfactory control mechanism for the disposal of low-level radioactive wastes'.[32] McEwan also pointed out that risks posed by nuclear-powered vessels docked in New Zealand ports were small in relation to those already posted by hazardous cargoes.

As well as outlawing the stationing or testing of nuclear weapons in the zone, the Act prevented New Zealand's Armed Forces from possessing or controlling nuclear weapons whether inside or outside the zone. While it outlawed visits from nuclear-powered ships, the Act did not restrict scientific applications of nuclear technology or nuclear power for electricity generation, although this would be a source of confusion as New Zealanders increasingly came to think of their country as 'nuclear-free'.

Later that year, Labour campaigned in the general election on its success in making New Zealand nuclear-free. Labour won the election, but its programme of economic reforms rapidly lost the party popularity. But as

NEW ZEALAND BANNED IT.
KEEP NEW ZEALAND NUCLEAR FREE.
VOTE LABOUR ON AUGUST 15.

The Labour Party's campaign poster for the 1987 general election played up the party's role in banning nuclear weapons from New Zealand land and waters. Author's collection.

David Lange later wrote, as 'the popularity of the Labour Government shrivelled', support for the nuclear-free legislation remained strong: 'nuclear-free New Zealand had taken on a life of its own'.[33]

NEW ZEALAND GETS NUKED, TOO

Legislating for a nuclear-free New Zealand was a stand against the nuclear arms race and for the protection of New Zealanders from direct nuclear strikes or nuclear accidents, but it didn't protect New Zealand from the impact of a northern hemisphere nuclear war. The United States and Soviet use of the nuclear threat throughout the Cold War has been described as 'racking the nerves of generations' and New Zealanders' nerves were no exception.[34] After a series of successful moves to control and limit nuclear arms in the 1960s and 1970s — the Partial Test Ban Treaty of 1963, the Non-Proliferation Treaty of 1968, the Anti-Ballistic Missile Treaty of 1972 — the early 1980s had seen a deterioration of relations between the United States and the Soviet Union. Arms control talks failed during Ronald Reagan's first term in office (1981–84), by which time the nuclear arsenals of the world contained tens of thousands of weapons.

Nuclear war was considered to be a very real possibility. The 1974 Pugwash International Conference on Science and Public Affairs had put the likelihood of nuclear war at 50:50, a view the Australian Office of National Assessment reiterated in 1981. New Zealanders tended to agree. Envisaging New Zealand in 2001, Robert Mann wrote in 1981 that he expected New Zealand 'only a couple of decades from now to be subjected to the effects of a major nuclear war'.[35] Twenty-four per cent of Aucklanders polled in 1982 by the Commission for the Future's Study Group on Nuclear Disaster thought there was an even chance of nuclear war in the next 20 years, with a further 34 per cent thinking it was likely there would be a nuclear war.

In the Commission for the Future report on Nuclear Disaster published in 1982, two local attack scenarios were considered: one, a 1-kiloton tactical warhead launched at a nuclear-powered vessel docked in Devonport; the other, a 1-megaton airburst. The first scenario envisaged:

A guided missile cruiser (eg. US 'Long Beach') berthed at the Devonport Naval Base, is attacked by a hostile vessel using a 1 kt tactical nuclear weapon. Lethal nuclear radiation (600 rem) reaches out to a distance of 800m, while extensive blast damage (5psi) occurs out to 450m. The core of the 430 MW reactor is vaporised and combines with the radioactivity derived from the weapon itself; both rise with the fireball and return to earth in the manner characteristic of fallout from the explosion of a weapon alone. The plutonium from the nuclear weapons carried on board adds to strike weapon and reactor fission products . . . The attack causes thousands, or a few tens of thousands, of civilian casualties.[36]

The authors pointed out that New Zealand was out of range of Soviet land-based systems and the attacks were therefore unlikely, but not implausible, over the intended 30-year lifetime of the report. In concluding, the authors of the report said that the most serious impact on New Zealand of a northern hemisphere war was unlikely to result from fallout or other weapons effects but from the loss of trading partners. They recommended that, rather than continuing to 'ignore the possibility of nuclear war', New Zealand should be planning to survive one.[37] The report was not well received by the National Government, with several Cabinet ministers describing it as 'vague' and 'emotive'. Just two months later, the Government abolished the New Zealand Commission for the Future, saying 'recent publications show that the Commission's work was no longer relevant to the issues facing New Zealand'.[38]

While New Zealand's Ministry of Civil Defence was set up in 1960 specifically to deal with the threat of nuclear war, by the 1980s it dealt almost exclusively with the threat of natural disasters, especially floods. In a 1983 interview with the *New Zealand Listener*, George Preddey, one of the authors of the Commission for the Future's report and now assistant director-general of the Ministry of Civil Defence, said the British attitude to nuclear civil defence — with its little pamphlets suggesting people put brown paper over their windows in the event of a nuclear disaster — might encourage morale, but it wasn't realistic. 'Our attitude here,' he said, 'is that it is quite misleading to suggest that there is any effective response to nuclear attack. We believe there is no effective civil defence response, that it is unrealistic to plan for a direct nuclear attack on this country.'[39]

While he continued to query whether or not there was a credible civil defence response to a nuclear attack, Preddey did have suggestions on how New Zealand could prepare for a northern hemisphere nuclear war, or an attack on Australia. In his 1985 book, *Nuclear Disaster: A New Way of Thinking Down Under*, Preddey suggested that civil defence preparations for a nuclear disaster could include:

> An infrastructure to co-ordinate the mobilisation of every element of New Zealand society and the economy in the event of a nuclear disaster;
>
> Deployment of emergency monitoring equipment (for fallout, ultraviolet light, acid rain, and other contingencies of nuclear war) and the training of personnel to use this;
>
> Distribution of appropriate emergency medical supplies, perhaps including potassium iodate tablets (to block iodine-131 uptake in the event of major attacks on Australia), sun filtering creams (to block ultraviolet light), eye protection, etc;

Dissemination of authoritative, accurate information for the public on the likely immediate hazards, essential if mass panic and the worst psychological impacts were to be avoided.[40]

Opinion polls taken as part of the 1986 Defence Review showed that New Zealanders thought the country should prepare or plan for the aftermath of a nuclear war in the northern hemisphere. Some 60 per cent of people thought New Zealand should be developing all possible plans for coping with post-war conditions, and a further 25 per cent of respondants thought the country should be preparating by building shelters and storing food and water.

Meanwhile, another threat was looming. Atmospheric chemists first suggested in 1982 that there was a possibility of climate change induced by nuclear war. They calculated that the fires that would burn for weeks after a nuclear war — from burning cities, croplands and forest, and stored fossil fuels — would produce a thick layer of smoke that would 'drastically reduce the amount of sunlight reaching the earth's surface'.[41] This would almost totally eliminate agricultural production in the northern hemisphere, meaning that no food would be available for any survivors of a nuclear war. A subsequent study, the first to use the phrase 'nuclear winter', found that a global nuclear war could lead to sub-freezing land temperatures in continental areas — down to minus 15–25°C — for many months.[42] Further studies supported the idea of a nuclear winter; as it would affect the southern as well as the northern hemisphere, the matter caught the attention of the New Zealand media and public, and on 21 October 1984 a group of scientists took part in a nuclear winter debate on TV1's *Sunday* programme. In his 1987 book *Beyond Darkness*, climate scientist Barrie Pittock warned that a northern hemisphere war could leave temperatures in New Zealand and Australia 5–10°C cooler, with rainfall reduced to less than 50 per cent of normal. Conditions would be worse in the northern hemisphere, though, with the cold, dry conditions having a devasting impact on agriculture and leading to 'mass starvation' in the most affected countries and an influx of 'nuclear refugees' to New Zealand and Australia.[43]

Other individuals and organisations issued their own books and pamphlets about nuclear safety. Brian Hildreth's *A Nuclear Survival Manual for New Zealanders*, published in 1986, outlined preparation and protection measures for surviving in the aftermath of a nuclear war, including survival first aid, energy and self-reliance. It wasn't a pleasant world that was envisaged:

One of the immediate consequences of a nuclear war will be the breakdown and probable collapse of the complex organisation of human society. As a survivor, you must be acutely aware at all times of the dangers this breakdown will produce. Part of your survival strategy must be to maintain constant vigilance against other human beings if your physical safety is to be ensured.

As well as plans to establish a hidden campsite (including a decoy site to foil would-be interlopers) and how to store and hide a food cache, the book included instructions for first aid, midwifery and dealing with 'survival stress'.[44]

A more moderate view was provided in *New Zealand After Nuclear War* in 1987. Funded by the Ministry for the Environment and published by the New Zealand Planning Council, the book was a national case study of the effect on New Zealand of a large-scale war in the northern hemisphere. Focused on the impact on food, health, energy, communications and transport, it updated the Commission for the Future's work by looking at nuclear winter and at the effects of an electromagnetic pulse, which it described as having the potential to cause far greater devastation to New Zealand than radioactive fallout or a nuclear winter. The study concluded that, while New Zealand was not likely to be a direct target or suffer direct effects of a nuclear war, the most serious long-term effects 'would be caused by the loss of imported supplies on which every sector of activity in New Zealand depends and the loss of markets for export production which shapes much of the social and economic structure of the country'.[45] The authors said New Zealand needed programmes to improve public knowledge of the likely impact of a nuclear war, contingency plans for action if war occurred, and action to reduce New Zealand's vulnerability to the effects of nuclear war.

But by 1987, not only was New Zealand now 'nuclear-free', but the international nuclear threat was diminishing. In December that year, presidents Ronald Reagan and Mikhail Gorbachev signed a historic agreement to eliminate all intermediate- and shorter-range nuclear forces, removing one-fifth of the nuclear weapons in the world. Two years later, in 1989, the Soviet empire collapsed and the Berlin Wall that had divided Communist East Germany and capitalist West Germany since 1961 was opened. By 1991, the Soviet Union had disintegrated and the United States was the only remaining global superpower.

CONCLUSION

★

Nuclear-free
New Zealand
Can we take it
for granted?

*We are no longer a colony, we are no longer hanging on to the skirts
of major powers, we are a nation on our own and we are prepared to
stand up and face the world on our own responsibility.*
— GERARD WALL, MP, 1972[1]

*In rightly trying to find all effective methods of opposing nuclear
armaments, we can easily come to feel that we must oppose all
things radioactive, thus wasting effort and putting ourselves in an
indefensible position.* — JIM McCAHON, 1988[2]

I t would be easy to think that New Zealand's nuclear-free legislation, and the
public opinion that it reflected, and then reinforced, reflects the courageous
and independent way that New Zealanders have always thought, but this
is only part of the story. As many other publications have shown, some New
Zealanders were quick to recognise the hazards of radiation and the moral
indefensibility of nuclear weapons. The 'anti-nuclear' movement began with
the first peace march in 1947 and evolved through decades of protest against
nuclear testing in the Pacific and, later, protests against visits from nuclear-
powered ships and against the introduction of nuclear power, culminating in

the introduction of the New Zealand Nuclear Free Zone, Disarmament and Arms Control Act 1987.

As this book shows, however, alongside this 'anti-nuclear' movement was what can be seen as a 'pro-nuclear' movement. Although they were not a coherent group, and did not form an organised lobby, New Zealand contained many people and organisations with a pro-nuclear attitude, and the country was subject to the influence of outside organisations — most importantly, the United Kingdom Atomic Energy Authority (UKAEA) and the United States Atomic Energy Commission (USAEC) — who had their own reasons for wanting New Zealand to adopt nuclear technology or engage in joint scientific projects.

Between these two groups, or voices — the anti-nuclear and the pro-nuclear — options were considered and decisions made. Again and again it can be seen that, rather than be pushed in one direction or another by groups or individuals with a particular opinion or agenda, the people with decision-making power made practical decisions based on economics and national interest when it came to deciding whether or not to adopt a certain piece of nuclear technology or whether or not to participate in projects or ventures with international agencies. The 1950s and 1960s saw the United States and the United Kingdom looking to establish in New Zealand a nuclear partner whose dependencies on the bigger nation would be to that nation's advantage. New Zealand officials, however, co-operated only in a self-interested way, moving forward in the field of nuclear science only to the extent that it would benefit the advancement of New Zealand's own scientific and economic goals.

While there was never any call for New Zealand to have its own nuclear weapons, in the 1940s, 1950s and 1960s New Zealanders embraced nuclear technology and were as excited about the dawning atomic age as any nation's people. In the first few decades of the twentieth century, New Zealand medics and scientists made great use of the discoveries of radium and x-rays, where the risks were relatively low but the benefits were very high. 'The public are mad on radium,' the government balneologist said in 1914, and it was the generosity of the New Zealand public — whose donations helped the hospitals to purchase radium for cancer treatment — that meant that by 1929 New Zealand had a greater supply of radium per head than the United Kingdom. The field of nuclear physics began with the discovery of the atomic nucleus by New Zealander Ernest Rutherford. While New

Zealand's association with Rutherford did not drive New Zealand's uptake of the new technologies, Rutherford did help educate the public on the new science in lectures on his visits to New Zealand. He also helped New Zealand scientists procure supplies of radium and, most importantly, he facilitated the appointment of his former student, Ernest Marsden, to positions first at Victoria University College and then as head of the Department of Scientific and Industrial Research (DSIR).

When the Second World War led to the search for a means to turn the energy known to reside in the atomic nucleus into a weapon, a group of young DSIR scientists was seconded to work with British teams on the North American project. New Zealand's support for the British nuclear programme in this way can be seen as an extension of the historical military and scientific association with the United Kingdom, but there were strong elements of self-interest there, too. New Zealand scientists had access to the American nuclear projects because of the direct intervention of Marsden, who hoped to use the American-trained scientists to set up a nuclear science team at the DSIR after the war. Once the bombs had been dropped on Japan, and New Zealand's involvement was revealed, newspapers reported proudly about the scientists who had worked on the North American nuclear projects during the war, and wrote hopefully about the search for uranium — the fuel for atomic weapons and power — that resumed in 1946, no longer as a secret.

Following the Second World War, there was promise of a new atomic age. There was a lot of enthusiasm, official and public, for nuclear power to provide electricity for New Zealand, and support for the search for uranium to boost the West Coast economy and provide a new export industry for New Zealand. Significant government and private resources were poured into this effort. The nuclear advocates promoting these ventures were more concerned with the perceived economic benefits of nuclear technologies than any issues of safety or waste management, and in the 1950s there was no organised opposition to nuclear power or the possibility of a uranium mining industry.

New Zealand entered two partnerships with the UKAEA in the 1950s. One, to produce heavy water for the UKAEA and electricity for New Zealand at a geothermal power plant at Wairakei; the second, whereby the UKAEA funded prospecting efforts on the West Coast in return for the first right of refusal over any uranium mined. There were also partnerships with the USAEC. A bilateral agreement signed in 1956 allowed for the exchange of information regarding the design, construction and operation of a research reactor and allowed for the lease of up to 6 kilograms of enriched uranium for use as reactor fuel. The agreement resulted in gifts worth more than US$2 million in today's terms, including a sub-critical nuclear reactor, being given to New Zealand universities and the DSIR. In other agreements, DSIR scientists collected samples and conducted research for the USAEC and the UKAEA as part of monitoring the effects of radioactive fallout from their bomb tests. I have found no record of any opposition to any of these agreements with two of the world's nuclear powers.

In all of these projects and partnerships, New Zealand officials took advantage of opportunities that would be in the country's — or their organisation's — best interest: for example, by accepting gifts of laboratory equipment from the USAEC, and entering agreements with the UKAEA that would assist the development of the Wairakei geothermal field and the Buller Gorge uranium deposits and lead to jobs to boost the local economies. But if proposals were not in New Zealand's interest, New Zealand said no. Officials at the DSIR and the State Hydro-electric Department refused or delayed opportunities to set up a nuclear reactor in New Zealand, even after repeated approaches from the United States. Having a nuclear reactor was not seen as necessary, and the opportunity cost was high; at a time of tight science budgets, its operating costs would take money away from what were seen as more useful projects.

Alongside these 'peaceful' uses of nuclear technology was the ever-present spectre of nuclear weapons tests in the Pacific, to which a segment of the New Zealand population was always opposed. The New Zealand Government initially supported the British nuclear-testing programme, which ran from 1952 until 1958, by providing logistical support, most significantly in the form of two frigates to act as weather ships for the 1957 and 1958 series of hydrogen bomb tests. Significantly, however, New Zealand Prime Minister Sidney Holland refused a 1955 British request to test these bombs on New Zealand territory: Holland feared public opinion would be against using the Kermadec Islands for the test, and was concerned that this could upset his narrow majority in Parliament.

Opposition to nuclear weapons testing grew during the 1950s and 1960s, fuelled by growing information about the levels of radioactive fallout being deposited in New Zealand and around the world. The 1962 American weapons test at Johnston Atoll upset radio-communications throughout the South Pacific and filled the skies above New Zealand with an eerie glow that

The New Zealand Herald described as doing 'more than a hundred protest marches to fill men's minds with dread'.[3] By the time France started testing nuclear weapons in the South Pacific, in 1966, the New Zealand Government was opposed to nuclear weapons testing, and, in striking contrast to the earlier support for the British tests, in 1973 New Zealand sent a protest frigate to Moruroa to protest against the French tests. Opposition to nuclear weapons testing was now firmly established, and had started to expand to include opposition to visits by warships carrying nuclear weapons or fuelled by nuclear power. Contrast the 1960 visit of the American nuclear submarine the USS *Halibut*, which was welcomed and marvelled at, with the colourful protests against American nuclear ship visits of the 1970s and 1980s, where people marched in the streets and ships were met by the Peace Squadron of protest boats.

Fuelled partly by opposition to nuclear-powered warships and concern about nuclear accidents and radiation leaks, antipathy to the idea of nuclear power began to grow. Most of the submissions to the Royal Commission into Nuclear Power Generation in New Zealand were against the introduction of nuclear power, some for economic reasons and others because of health and safety concerns about the risks of nuclear power. The New Zealand Electricity Department had had nuclear power on the national power plan from 1964, but, by the time the commission released its report in 1978, new indigenous fuel sources — gas and coal — had been found and electricity demand forecasts had been greatly reduced, so the question of whether or not to start building a nuclear power station did not need to be put to a full ideological or economic test. But at the same time as nuclear power was being considered, New Zealand and international companies were continuing to prospect for uranium on the West Coast with no political or grassroots opposition. In the 1990s uranium prospecting was made unlawful in New Zealand, probably as a reflection of the public's broad interpretation of what it means for New Zealand to be 'nuclear-free'. In the 1970s, however, the possibility of a uranium mining industry was not seen as a bad thing.

When New Zealand's nuclear-free legislation was introduced in 1987 it applied to nuclear weapons, nuclear-powered ships and nuclear waste. By then, New Zealand was far-sighted enough to be considering not just national interests but also global interests, and was making a point about the folly of nuclear weapons. New Zealanders were immediately on the world stage, and lauded for being independent and courageous. In the years that followed, this nuclear-free ethos became deeply entrenched, a formidable part of national identity, that spread to nuclear power, to uranium prospecting and often to anything associated with nuclear technology and radiation. Like other decisions in the country's nuclear history, New Zealand's rejection of any involvement with nuclear weapons could be seen as a continuation of a series of pragmatic and self-interested decisions. Nuclear weapons are in no one's best interest, and New Zealand was independent enough and bold enough to say so. But it was only a few years before the nuclear-free legislation was introduced that New Zealand had rejected nuclear power for electricity generation (for the time being). Companies had tired of spending money on attempting to prove the uranium deposits on the West Coast, and had abandoned their prospecting camps and drilling projects. The nuclear-free legislation, which was focused on nuclear weapons, therefore hit a New Zealand with limited so-called 'peaceful' uses of nuclear technology — no nuclear power, no uranium mining, no research reactors — and the anti-nuclear attitude was able to spread, unchallenged, beyond the initial logical antipathy to nuclear weapons. This may seem obvious, but what is not so well known is that, as this book shows, New Zealand's limited uses of peaceful nuclear technology by the 1980s was not because of any ideological reasons, but because of a series of economic decisions.

New Zealand's 'rejection' of nuclear power and uranium mining can now be seen as the consequence of economic decisions made in the 1970s. The fact that we do not have a nuclear reactor can be seen as a pragmatic decision made by a country with a limited budget for science that it chose to focus on agriculture and supporting sciences, and the lack of a heavy-water plant can be seen as the result of the British withdrawing from a planned joint venture. The fact that economic arguments were used to argue *against* these things makes it possible that economic arguments could be used to argue *for* these things — nuclear power and uranium mining in particular — in the future.

New Zealand is 'nuclear-free' when it comes to nuclear weapons, nuclear power and uranium mining. But given the pattern of history, with its nuclear advocates as well as anti-nuclear lobbyists, and in light of recent calls for aspects of New Zealand's nuclear-free policy to be reviewed, New Zealand's nuclear-free policy — in the broad sense in which it is currently interpreted — cannot be taken for granted.

NOTES

Preface

1 From Prime Minister Helen Clark's address on 'New Zealand and Peaceful Conflict Resolution' in Cairo, Egypt, www.beehive.govt.nz/speech/new+zeal and+and+peaceful+conflict+resolution, downloaded 23 Jun. 2010.
2 A statement made when presenting his New Zealand Nuclear Free Zone, Disarmament, and Arms Control (Nuclear Propulsion Reform) Amendment Bill to Parliament, www.parliament. nz/mi-NZ/PB/Debates/Debates/7/ c/1/47HansD_20050727_00001577-New-Zealand-Nuclear-Free-Zone-Disarmament.htm, downloaded 23 Jul. 2008.
3 NZPA, 'Reactor Tempers Hutt City's Nuke Free Declaration', www.rsnz. govt.nz/archives/news_pre_oct99/ news/y_1995/m _09/d_18/a_3.php, downloaded 12 Jan. 2000.
4 The Topp Twins interview, 'Nine to Noon', 31 Mar. 2009, www. radionz.co.nz/national/programmes/ ninetonoon/20090331, downloaded 21 Mar. 2010.

1   The public are mad on radium!

1 Wohlmann to General Manager of Department of Tourist and Health Resorts, 25 May 1914, TO1, 24/34, Archives New Zealand (ANZ).
2 *New York Herald Tribune*, 12 Sep. 1933. News clipping reproduced in Charles Weiner, 'Physics in the Great Depression', *Physics Today*, 23(10), 1970, p. 33.
3 Arthur S. Wohlmann, *Mineral Waters and Spas of New Zealand*, Government Printer, Wellington, 1914, p. 59; Wohlmann, *op cit.*
4 Wohlmann, *Mineral Waters, ibid.*, p. 61.
5 *Rotorua Times*, 30 Oct. 1914, TO1, 24/34, ANZ.

6 Wohlmann, *Mineral Waters, op cit.*, p. 62.
7 *New Zealand Mail*, 6 Feb. 1896, p. 36.
8 *New Zealand Medical Journal*, 9, Jul. 1896, p. 169.
9 W. T. L. Travers, 'Presidential Address', *Transactions and Proceedings of the New Zealand Institute*, 29, 1896, pp. 111–29, at p. 118.
10 James Ryan, Keith Sutton and Malcolm Baigent, *Australasian Radiology: A History*, McGraw-Hill, Sydney, 1996, p. 20.
11 C. C. Anderson, 'The Development of Radiology in New Zealand', *Australasian Radiology*, 10, 1966, p. 296.
12 Harry A. de Lautour, 'On the Localisation of Foreign Bodies by Means of the X-rays', *New Zealand Medical Journal*, 1(2), 1900, pp. 67–74, at p. 74.
13 Ernest Rutherford, 'Uranium Radiation and the Electrical Conduction Produced By It', first published in *Philosophical Magazine*, Jan. 1899; reproduced in Ernest Rutherford, *The Collected Papers of Lord Rutherford of Nelson, vol. 1*, George Allen & Unwin, London, 1962, pp. 169–215, at p. 175.
14 J. F. Borrie, 'A History of Training for the Diploma of the Society of Radiographers (London)', *Shadows*, 16(1), 1973, pp. 15–27, at p. 16.
15 Annual Report of the Director-General of Health, *Appendices to the Journals of the House of Representatives*, 1926, vol. 2, H-31, p. 29.
16 *The Press*, 19 Sep. 1902, p. 5.
17 *Ibid.*
18 *Ibid.*
19 *The Press*, 4 May 1914, p. 6.
20 *The Press*, 5 Dec. 1924, p. 14.
21 Letter from Fenwick to Department of Health, 6 May 1926, H1 15/1/6 (8486), ANZ.
22 Wohlmann, 25 May 1914, *op cit.*
23 *The Press*, 20 Jun. 1900, p. 6.
24 *The Dominion*, 29 Jan. 1930, p. 10.
25 C. C. Anderson, 'Results of X-ray

and Radium Therapy at the Dunedin Hospital', *New Zealand Medical Journal*, 28 (146), 1929, pp. 200–11, at p. 206.

26 Memo from General Manager to Duncan, 21 June 1922, TO1, 24/34, ANZ.

27 Memo from Tourist Agent to General Manager, 24 Sep. 1925, TO1, 24/34, ANZ.

28 From Circular Letter No. 21 Hosp 21/1931, H1, 131/1/6, ANZ.

29 *The Press*, 29 Jul. 1905, p. 10.

30 John Campbell, *Rutherford: Scientist Supreme*, Christchurch, AAS Publications, 1999, p. 291.

31 W. A. Knox, 'Protective Measures in Dental Roentgenology', *New Zealand Dental Journal*, Mar. 1920, pp. 137–9, at p. 137.

32 Anderson, 'The Development of Radiology in New Zealand', *op cit.*, p. 303.

33 John L. Campbell, 'Thirty-six Years in Radiography', *Health and Service*, 8(1), 1953, pp. 49–52, at p. 49.

34 Quoted in J. Samuel Walker, *Permissible Dose: A History of Radiation Protection in the Twentieth Century*, University of California Press, Berkeley, 2000, p. 5.

35 *The Evening Post*, 21 Jan. 1933, p. 11.

36 *Ibid.*; *Barrier Miner*, 2 April 1932, p. 1.

37 Charles F. Hines, 'The Use and Care of Radium in Hospital Practice', *New Zealand Nursing Journal*, 34(12), 1941, pp. 400–4, at p. 403.

38 Annual Report for the DSIR, *Appendices to the Journal of the House of Representatives* 1937–38, vol. 3, H-34, p. 88.

39 Wohlmann, 25 May 1914, *op cit.*

40 *New Zealand Journal of Health and Hospitals*, 4(9), 1921, pp. 249–51 at pp. 249, 251, H1, 53/19 (28298), ANZ.

41 *Ibid.*, p. 249.

42 Campbell, 'Thirty-six Years in Radiography', *op cit.*, p. 49.

43 Annual Report of the Department of Health, *Appendices to the Journal of the House of Representatives* 1921–22, vol. 3, H-31, p. 19.

44 Circular letter from Director-General of Health to medical superintendents of all hospitals, 14 Aug. 1933, H1 53/19 (28298), ANZ.

45 'Recommendations of the X-Ray and Radium Protection Committee', *Journal of the Hospital Boards' Association of New Zealand*, Sept. 1933, pp. 24–6, at p. 24.

46 Ryan *et al.*, *op cit.*, p. 339.

47 G. E. Roth, 'The Physical Services to Radiology in New Zealand', *New Zealand Medical Journal*, 45(248), 1946, pp. 384–8, at p. 387.

48 Ernest Rutherford, 'The Development of the Theory of Atomic Structure', in Joseph Needham and Walter Pagel, *Background to Modern Science*, Cambridge University Press, Cambridge, 1938, pp. 61–74, at p. 68.

49 E. Rutherford and F. Soddy, 'Radioactive Change'. First published in *Philosophical Magazine*, May 1903, reproduced in Rutherford, *The Collected Papers, vol. 1, op cit.*, pp. 596–608, at p. 608.

50 A. S. Eve, *Rutherford: Being the Life and Letters of the Rt Hon Lord Rutherford, O. M.*, Cambridge University Press, Cambridge, 1939, p. 102.

51 *The New Zealand Herald*, 31 Aug. 1914, p. 8.

52 Brian Cathcart, *The Fly in the Cathedral: How a Small Group of Cambridge Scientists Won the Race to Split the Atom*, Viking, London, 2004, pp. 247, 249.

53 *New York Herald Tribune*, 12 Sep. 1933, news clipping reproduced in Charles Weiner, 'Physics in the Great Depression', *Physics Today*, 23(10), 1970, p. 33.

2 Some fool in a laboratory

1 Viscount Bledisloe to Minister of External Affairs, 9 Aug. 1945, EA1, W2619, 121/1/1, part 1, ANZ.

2 Kate Dewes and Robert Green, *Aotearoa/New Zealand at the World Court*, The Raven Press, Christchurch, 1999, p. 9; Glyn Strange, 'Popper's A-bomb dilemma', *The Evening Post*, 14 Aug. 1995, p. 5.

3 'New Zealand Participation in Atomic Bomb Development', issued to the press on 13 Aug. 1945, EA1, W2619, 121/1/1, part 1, ANZ.

4 W. David McIntyre, 'From Dual Dependency to Nuclear Free', in Geoffrey W. Rice (ed.), *The Oxford History of New Zealand*, 2nd edn, Oxford University Press, Auckland, 1992, pp. 520–38, at p. 524.

5 T. R. Ritchie to Marsden, 29 Sep. 1943, H1, 108/7/1, 45543, 1944–61, ANZ.

6 Marsden to Ritchie, 4 Nov. 1943, H1, 108/7/1, 45543, 1944–61, ANZ.

7 Director-General of Health to Captain Chisholm, 23 May 1944, H1, 108/7/1, 45543, 1944–61, ANZ.

8 *Ibid.*

9 George Roth, 'Radon Micro Determination by the Curtiss-Davis α-particle Counting Method', *New Zealand Journal of Science & Technology* 27B(2), 1945, pp. 147–53, at p. 152.

10 Margaret Gowing, *Britain and Atomic Energy 1939–1945*, Macmillan, London, 1964, p. 394.

11 M. G. Oliphant, in Marsden Editorial Committee, *Sir Ernest Marsden 80th Birthday Book*, A. H. & A. W. Reed, Wellington, 1969, p. 102.

12 H. H. Massey, in *ibid.*, p. 47.

13 Robin Williams, 'Reflections on My Involvement in the Manhattan Project', seminar at Victoria University of Wellington, 10 Aug. 2001.

14 Gowing, *op cit.*, p. 258.

15 Williams, *op cit.*

16 Gowing, *op cit.*, p. 276.

17 Scientific Liaison Officer to Marsden, 4 Jan. 1945, SIR1, W1414, 74/10, quoted in 'World War II narrative No. 9', AAOQ, W3424 (box 16), ANZ.

18 Marsden to Minister of Scientific and Industrial Research, 30 Jan. 1945, SIR1, W1414, 74/10, ANZ.

19 Marsden to Cockcroft, 26 Mar. 1945, SIR1, W1414, 74/10, ANZ.

20 Marsden to Watson-Munro, 12 Apr. 1945, SIR1, W1414, 74/10, ANZ.

21 Marsden to George, 5 Apr. 1945, SIR1, W1414, 74/10, ANZ.

22 C. R. Barnicoat, in *Sir Ernest Marsden 80th Birthday Book, op cit.*, p. 3; and F. J. Brogan, in *ibid.*, p. 57.

23 Marsden to Hon. D. G. Sullivan, 6 Jul. 1944, SIR1, W1414, 74/10, ANZ.

24 Personal recollections by Jim McCahon, recorded by McCahon over 14–16 Aug. 1998 as notes for talk to Kaikoura Probus Club.

25 Richard Rhodes, *The Making of the Atomic Bomb*, Touchstone, New York, 1986, p. 735.

26 Editorial, 'Horror with Some Hope', *New Zealand Listener*, 17 Aug. 1945, p. 5.

27 Dewes and Green, *op cit.*, p. 9; Strange, *op cit.*

28 Page to Marsden, 19 Sep. 1944, SIR1, W1414, 74/10, part 1, ANZ.

29 Personal recollections by Jim McCahon, *op cit.*

30 *New Zealand Parliamentary Debates* 269, 1945, p. 266.

31 *Ibid.*, p. 752.

32 *Ibid.*, p. 486.

33 Letter to Attlee, signed by Williams and others, 19 Sep. 1945, Robin Williams's personal archives.

34 Marsden to Minister of Scientific and Industrial Research, 12 Sep. 1945, EA1, W2619, 121/1/1, part 1, ANZ.

35 'World War II narrative No. 9', *op cit.*

36 Laurie Brocklebank, *Jayforce: New Zealand and the Military Occupation of Japan 1945–48*, Oxford University Press, Auckland, 1997, p. 121.

37 Marsden to Minister of Scientific and Industrial Research, 9 Aug. 1945, SIR1, W1414, 74/10, ANZ.

38 K. T. Fowler, 'The "New Golden Hind" Looks for Uranium', *The Public Service Journal*, May 1946, pp. 196–9, at p. 197.

39 Marsden to Appleton, 29 Sep. 1945, SIR1, W1414, 74/10, quoted in 'World War II narrative No. 9', *op cit.*

40 *Ibid.*

41 Quoted in Marsden to Acting Minister of Scientific and Industrial Research, 13 May 1947, SIR1, W1414, 74/10, ANZ.

42 Draft memorandum by Marsden, 18 Sep. 1947, AAOQ, W3424, 74/20/- (1947–55), ANZ.

43 Shanahan to Ushwin, 18 Oct. 1948, EA, W2619, 121/2/1, ANZ.

44 Watson-Munro to Shanahan, 17 Sep. 1948, EA, W2619, 121/2/1, part 1, ANZ.

45 E. Marsden, 'The Atomic Age', unpublished talk to National Dairy Association of New Zealand in 1947, MS-Papers-1342-269, Alexander Turnbull Library (ATL).

46 Marsden to Taylor, 17 Jan. 1955, MS-Papers 1342-37, ATL.

47 Marsden to Shanahan, 11 Dec. 1947, AAOQ, W3424, 74/20/- (1948–49), ANZ.

48 Marsden and Watson-Munro, 'An Atomic Pile for New Zealand', EA1, W2619, 121/2/1, part 1, ANZ.

49 Watson-Munro, 'Amplification of Reasons for an Atomic Pile in New Zealand', EA, W2619, 121/2/1, part 1, ANZ.

50 Clement Attlee to Peter Fraser, 30 Sep. 1948, EA, W2619, 121/2/1, part 1, ANZ.

51 Notes for Prime Minister in Connection with Visit to Harwell, 17 Dec. 1948, EA, W2619, 121/2/1, part 1, ANZ.

3 Cold War and red-hot science

1 This recollection is of the last Operation Grapple blast on 23 September 1958, from Maurice Hayman, *Those Useless Wings: Operation Grapple*, M. Hayman, Riverton, 1997, p. 13.
2 This Department of Health report is referred to in a letter from G. R. Laking, Secretary of External Affairs to Secretary of the Department of Island Territories, 15 Apr. 1957, H1, 26717, 108/11, 1951–57, ANZ.
3 Stewart Firth, *Nuclear Playground*, Allen & Unwin, Sydney, 1987.
4 W. David McIntyre, 'From Dual Dependency to Nuclear Free', in Geoffrey W. Rice (ed.), *The Oxford History of New Zealand*, 2nd edn, Oxford University Press, Auckland, 1992, pp. 520–38.
5 Editorial, *New Zealand Listener*, 12 Jul. 1946, p. 5.
6 *New Zealand Parliamentary Debates*, 273, 1946, p. 224.
7 James Robert McClelland, *Report of the Royal Commission into British Nuclear Tests in Australia, vol. 1*, Australian Government Publishing Service, Canberra, 1985, p. 15.
8 Wayne Reynolds, *Australia's Bid for the Atomic Bomb*, Melbourne University Press, Melbourne, 2000.
9 *The Dominion*, 6 Oct. 1952, p. 7.
10 *The Dominion*, 7 Oct. 1952, p. 6.
11 *The Press*, 16 Oct. 1953, p. 11.
12 *The Dominion*, 15 May 2001, p. 3.
13 *Ibid*.
14 Lawrence Badash, *Scientists and the Development of Nuclear Weapons: From Fission to the Limited Test Ban Treaty 1939–1963*, Humanity Books, New York, 1995, p. 83.
15 *New Zealand Parliamentary Debates*, 303, 1954, pp. 198–9.
16 Atomic and Thermo-Nuclear Tests in the Pacific, 6 April 1954, ABHS 950, W4627/3544, 121/5/2, part 1, ANZ.
17 Quoted in Malcolm Templeton, *Standing Upright Here: New Zealand in the Nuclear Age 1944–1990*, Victoria University Press, Wellington, 2006, pp. 65–6.
18 Minutes of the 16th Meeting of the Defence Science (Policy) Committee, 9 October 1952, AAOQ, W3424 (box 5), 74/22/-, vol. 1(b), ANZ.
19 Geoffrey Scoones to Sidney Holland, 11 Dec. 1953, ABHS 950, W5422 (box 166), 111/18/3/1, part 1, ANZ.
20 Marsden to Hamilton, 24 Mar. 1954, AAOQ, W3424 (box 5), 74/22/-, part 3, ANZ.
21 *The Daily Telegraph*, 23 Feb. 1955, p. 8.
22 Kevin Clements, *Back from the Brink: The Creation of a Nuclear-Free New Zealand*, Allen & Unwin/Port Nicholson Press, Wellington, 1988, p. 32.
23 *New Zealand Parliamentary Debates*, 303, 1954, p. 642.
24 Press statement distributed by External Affairs, 26 Jan. 1956, ABHS 950, W4627, 121/2/2, part 2, ANZ.
25 Cockcroft to Marsden, 24 Jan. 1956, ABHS 950, W4627, 121/2/2, part 2, ANZ.
26 Lorna Arnold, *Britain and the H-bomb*, Palgrave Macmillan, Houndmills, 2001, p. 95.
27 Holland to Scoones, High Commissioner for the United Kingdom, 6 Apr. 1954, ABHS 950, W4627 (box 3544), 121/5/2, part 2, ANZ.
28 Barry Gustafson, 'Holland, Sidney George 1893–1961', *Dictionary of New Zealand Biography*, www.dnzb.govt.nz, downloaded 8 Mar. 2010.
29 Eden to Holland, 18 May 1955, ABHS 950, W4627 (box 3544), 121/5/2, part 2, ANZ.
30 Message quoted in Scoones to Holland, 5 Jul. 1955, ABHS 950, W4627, 121/5/2, part 2, ANZ.
31 Marsden to Holland, 10 Jul. 1955, ABHS 950, W4627 (box 3544), 121/5/2, part 2, ANZ.
32 John Crawford, ' "A Political H-Bomb": New Zealand and the British Thermonuclear Weapon Tests of 1957–58', *The Journal of Imperial and Commonwealth History*, 26(1), 1998, pp. 127–50, at p. 133.
33 Eden to Holland, 2 Sep. 1955, ABHS 950, W4627 (box 3544), 121/5/2, part 2, ANZ.
34 *The Evening Post*, 2 Apr. 1956 (responding to report in *The Daily Express*), ABHS 950, W4627 (box 3544), 121/5/2, part 2, ANZ.
35 *New Zealand Parliamentary Debates* 308, 1956, p. 575.
36 Press statement, 28 May 1956, ABHS 950, W4627 (box 3544), 121/5/2, part 2, ANZ.
37 John Crawford, *The Involvement of the Royal New Zealand Navy in the British Nuclear Testing Programmes of 1957 and 1958*, New Zealand Defence Force, Wellington, 1989, p. 18.

38 Hayman, *op cit.*, p. 6.
39 *Here & Now* 59, 1957, p. 5.
40 Turbott to Laking, Secretary of External Affairs, 10 Apr. 1957, H1, 26717, 108/11, 1951–57, ANZ.
41 This Department of Health report is referred to in a letter from G. R. Laking, Secretary of External Affairs to Secretary of the Department of Island Territories, 15 Apr. 1957, *op cit.*
42 *The Dominion,* 17 May 1957, p. 10.
43 Templeton, *op cit.*, p. 79.
44 Quoted in Crawford, *The Involvement of the Royal New Zealand Navy, op cit.*, p. 33.
45 *The Dominion,* 17 May 1957, p. 8.
46 Sidney Holland, 'Nuclear Test Explosions', *External Affairs Review,* 7(5), 1957, pp. 17–20, at pp. 17, 19.
47 *The Press,* 10 Nov. 1997, p. 15.
48 Crawford, *The Involvement of the Royal New Zealand Navy, op cit.*, p. 31.
49 This recollection is of the last Operation Grapple blast on 23 September 1958, from Hayman, *op cit.*, p. 13.
50 *Ibid.*, p. 18.
51 Templeton, *op cit.*, p. 82.
52 *Ibid.*, p. 83.
53 *Ibid.*, p. 89.
54 Elsie Locke, *Peace People: A History of Peace Activities in New Zealand,* Hazard Press, Christchurch, 1992, p. 135.
55 *New Zealand Parliamentary Debates,* 291, 1950, p. 2144.
56 Monte Holcroft, 'The cloud that grew', *New Zealand Listener,* 11 Apr. 1958, p. 10.
57 Quoted in Templeton, *op cit.*, p. 67.
58 Holland, 'Nuclear Test Explosions', *op cit.*, p. 18.
59 *Ibid.*, pp. 19, 20.
60 John Lewis Gaddis, *The Cold War,* Allen Lane, London, 2005, p. 70.
61 *The New Zealand Herald,* 10 Jul. 1962, p. 1.
62 *The New Zealand Herald,* 11 Jul. 1962, p. 6.
63 David Lange, *Nuclear Free — The New Zealand Way,* Penguin, Auckland, 1990, p. 11.
64 *The New Zealand Herald,* 11 Jul. 1962, p. 6.
65 Templeton, *op cit.*, p. 100.
66 Ethel W. Wood, in *Sir Ernest Marsden 80th Birthday Book, op cit.*, p. 25.
67 *Auckland Star,* 8 May 1959, MS-Papers-1342-386, ATL.
68 *The Evening Post,* 6 Jun. 1962, MS-Papers-1342-379, ATL.
69 Marsden to Holyoake, 14 Jan. 1963, MS-Papers-1342-016, ATL.
70 Newspaper clipping, n.d., but *c.* 1964, and paper not identified, MS-Papers-1342-387, ATL.
71 *New Zealand Parliamentary Debates,* 337, 1963, p. 2696.
72 Newspaper clipping, publication not identified, n.d., but possibly 1961, MS-Papers-1342-387, ATL.
73 'Nuclear Testing Potential Danger', *Salient,* 29 Jun. 1965, p. 14, MS-Papers-1342-386, ATL.

4   Uranium fever!

1   *The Press,* 10 Nov. 1955, p. 12.
2   *The Press,* 15 Nov. 1955, p. 12.
3   *The Press,* 10 Nov. 1955, *op cit.*
4   *The Press,* 15 Nov. 1955, *op cit.*
5   *Ibid.*
6   Charles Roggi to G. R. Laking, 19 Apr. 1954, EA, W2619, 121/2/1, part 1, ANZ.
7   *The Press,* 15 Nov. 1955, p. 9.
8   *The Press,* 17 Nov. 1955, p. 12.
9   *Ibid.*
10 *The Press,* 24 Nov. 1955, p. 14.
11 *Ibid.*
12 *Ibid.*
13 R. W. Willett to Secretary of DSIR, 26 Jul. 1956, AATJ 6090, W4897/102, 5/22, part 3, ANZ.
14 Minister of Mines to F. Cassin, 20 Apr. 1956, CABH 3391, CH58, 6/27, ANZ CHCH.
15 E. H. Brooker to Minister of Mines, 25 Jun. 1957, CABH 3391, CH58, 6/27, 1955–62, ANZ CHCH.
16 *The Press,* 16 Aug. 1956, p. 10.
17 *The Press,* 31 Jan. 1957, p. 13.
18 *The Dominion,* 6 Jun. 1957, p. 12.
19 Lime and Marble to Goldfields and Mines Committee, 16 Jul. 1957, AATJ 6090, 3/38, part 3, ANZ.
20 Radio broadcast transcript reproduced in E. H. Brooker to Minister of Mines, 25 Jun. 1957, CABH 3391, CH58, 6/27, 1955–62, ANZ.
21 *Greymouth Evening Star,* 29 Jan. 1958, CABH 3391, CH58, 7/134, 1958–1982, part 3, ANZ CHCH.
22 *The New Zealand Herald,* 7 Sep. 1960, AATJ 6090, W4993, 23/2/1220/1, ANZ.
23 *Greymouth Evening Star,* 18 Mar. 1960,

in AATJ 6090, W4897/102, 5/22, part 3, ANZ.

24 Minister of Mines to Dr Gordon J. Williams, 4 May 1960, in AATJ 6090, W4897/102, 5/22, part 3, ANZ.

25 Proposal for a Regional Exploration and Prospecting Plan for Buller/Paparoa Uranium Province, 6 May 1960, in AATJ 6090, W4897/102, 5/22, part 3, ANZ.

26 *The Press*, 18 Aug. 1966, AATJ 6090, W4897/102, 5/22, part 4, ANZ.

27 B. D. P. Williamson, NRL, to Lime and Marble, 31 Oct. 1968, AATJ 6090, W5152/129, 12/46/1051, ANZ.

28 *Greymouth Evening Star*, 29 Jun. 1972, AATJ 6090, W4897/102, 5/22, part 5, ANZ.

29 Treasury report on Cabinet Paper dated 10 Jun. 1977, AATJ 6090, W4897/103, 5/22-1, part 3, ANZ.

5 There's strontium-90 in my milk

1 George Roth, *Radiation Hazards: A Survey*, Department of Health, Wellington, 1952, p. 1; H1, 26717, 108/11, 1951–57, ANZ.

2 George Roth, 'Radiation Hazards in Perspective', *New Zealand Science Review*, 23(1), 1965, pp. 8–13, at p. 8.

3 Recollection from Ruth Brassington, 2005. Written for the author.

4 Department of Health, Recommendations for Protection from Radiation Hazards, August 1951, H1, 108/11, 1951–57, ANZ.

5 Roth, *Radiation Hazards*, *op cit.*

6 G. E. Roth, 'The Problem of Radiation Safety', *The New Zealand Radiographer*, 1(4), 1949, pp. 3–14, at p. 4.

7 Roth, *Radiation Hazards*, *op cit.*, p. 3.

8 Turbott to Minister of Health, 16 May 1957, H1, 26717, 108/11, 1951–57, ANZ.

9 From file record of conversation between G. E. Roth and *The New Zealand Herald*, 22 Apr. 1958, H1, 26758, 108/11, 1957–58, ANZ.

10 Department of Health memo 1969/189, issued by B. W. Christmas, 26 Aug. 1969, CABI, CH91, 13/704, 1953–71, ANZ CHCH.

11 *Health*, 24(3), 1972, p. 16.

12 From a newspaper report in a Sydney newspaper in H1, 26717, 108/11, 1951–57, ANZ.

13 Department of Health memo 1971/147,

issued by Dr M. C. Laing, 12 Jul. 1971, CABI, CH91, 13/704, 1953–71, ANZ CHCH.

14 *The Dominion*, 13 Jul. 1957; H1, 26717, 108/11, 1951–57, ANZ.

15 *Ibid.*

16 Roth, *Radiation Hazards, op cit.*, p. 22.

17 Linus Pauling, 'Atomic Power and Radiation Hazards', *New Zealand Listener*, 18 Dec. 1959, pp. 6–7, at p. 7.

18 Report to United Nations Scientific Committee on the Effects of Atomic Radiation, 17 Feb. 1958, H1, 26758, 108/11 1957–58, ANZ.

19 Andrew McEwan, *Radiation Protection and Dosimetry in New Zealand: A History of the National Radiation Laboratory*, New Zealand Department of Health, Wellington, 1983.

20 Annual Report: AEC Contract AT(30-1)-2167, 13 May 1959, SIR1, W1414, 74/27/6, ANZ.

21 Hamilton to Minister of Scientific and Industrial Research, 24 Jan. 1963, SIR1, 74/38, ANZ.

22 Philippa Mein Smith, 'New Zealand Milk for "Building Britons" ', in Mary P. Sutphen (eds) and Bridie Andrews, *Medicine and Colonial Identity*, Rutledge, London, 2003, pp. 79–102, at p. 81.

23 'New Zealand Strontium-90 Figures "A warning" ', *New Zealand Public Service Journal*, Dec. 1959, p. 8.

24 H. J. Yeabsley, 'No Danger in N.Z. from Atomic Fallout', *Parent and Child*, 3(5), 1960, p. 26.

25 B. D. P. Williamson, 'Luminous Paint: No Significant Hazards Detected', *N. Z. Horological Journal*, 14(4), 1960, pp. 6–9, at p. 7.

26 Advertising flyer, H1, 26717, 108/11, 1951–57, ANZ.

27 G. E. Roth to Messers Claude W. Batten and Co., 7 Apr. 1954, H1, 26717, 108/11, 1951–57, ANZ.

28 *The Dominion*, 17 Apr. 1962, MS-Papers-1342-381, ATL.

29 *Manawatu Daily Times*, 10 Jan. 1955, p. 6.

30 J. V. Scott, Report on the Second United Nations Conference on the Peaceful Uses of Atomic Energy, p. 17, ED1, W2673, 2/0/22/5, part 2, ANZ.

31 'Hush-Up Over Deadly Cargo', *The Christchurch Star*, 20 Sep. 1975, p. 1.

32 *The Evening Post*, 22 Sep. 1975, ABQU 632, W4452, 108/5/6, ANZ.

33 Rachel Barrowman, *Victoria University*

of Wellington, 1899–1999: A History, Victoria University Press, Wellington, 1999.

34 *The Dominion,* 3 Jul. 1963, p. 1.

35 *The Dominion,* 5 Jul. 1963, p. 2.

36 *Weekend Star,* 9 Apr. 1988, pp. 1–2.

37 Suggested Organization of Radiological Monitoring and Recontamination Services for the New Zealand Civil Defence Organization, submitted to the Radiological Advisory Council at its meeting on 3 Dec. 1958, H1, 26758, 108/11, 1957–58, ANZ.

38 Ministry of Civil Defence, *Civil Defence in New Zealand,* Department of Internal Affairs, Wellington, 1959, pp. 6, 13.

39 From Roth's answers to a questionnaire sent by the Engineering Institute of the University of Michigan, H1, 26758, 108/11, 1957–58, ANZ.

40 McCahon to External Affairs, 6 February 1967, CABI, CH56, 23/20/1, 1967–68, ANZ CHCH.

41 *Canta,* 25 Jun. 1968, pp. 1–2, University of Canterbury Archives 16578, MB 1281, Box 2, Item 1/12, Newspaper cuttings 1944–67.

42 *Ibid.,* p. 3.

43 McEwan, *op cit.,* p. 64.

6   Atoms for Peace

1 J. Williams, 'Report on Development of Nuclear Sciences in New Zealand', 19 Jul. 1956. Quoted in Barrowman, *op cit.,* p. 166.

2 *The Dominion,* 17 Oct. 1956, ED1, W2673, 2/0/65/5, ANZ.

3 USS *Halibut* Nuclear Submarine, www. nzhistory.net.nz/media/video/uss-halibut-nuclear-submarine, downloaded 20 Apr. 2010.

4 *The New Zealand Herald,* 20 Apr. 1960, p. 13.

5 Athol Rafter, extract from 'Problems in the Establishment of a Carbon-14 and Tritium Laboratory', paper presented at the Sixth International Conference on Radiocarbon and Tritium Dating, Washington State University, 1965, in Rebecca Priestley (ed.), *The Awa Book of New Zealand Science,* Awa Press, Wellington, 2008, pp. 250–4, at p. 251.

6 *Ibid.*

7 *The Evening Post,* 6 Oct. 1952, EA1, W2619, 121/2/1, part 1, ANZ.

8 Address by Mr Dwight D. Eisenhower, President of the United States of America, to the 470th Plenary Meeting of the United Nations General Assembly, www. iaea.org/About/history_speech.html, downloaded 15 Sep. 2009.

9 Speech by Munro to UN General Assembly, *External Affairs Review,* 4(11), 1954, p. 9.

10 A. E. Davenport to Secretary for External Affairs, 30 May 1955, ED1, W2673, 2/0/22/5, ANZ.

11 Secretary of DSIR to Secretary of External Affairs, 5 Apr. 1955, ED1, W2673, 2/0/22/5, ANZ.

12 Secretary of DSIR to Secretary of External Affairs, 14 Jun. 1955, ED1, W2673, 2/0/22/5, part 1, ANZ.

13 *External Affairs Review,* 6(6), 1956, pp. 8–9.

14 Lloyd White, Memo on International Conference on the Peaceful Uses of Atomic Energy, Geneva, 1955, ED1, W2673, 2/0/22/5, ANZ.

15 Templeton, *op cit.,* p. 100.

16 W. M. Hamilton, Atomic Energy (report on 1955 overseas visit), n.d. but *c.* 1955, ED1, W2673, 2/0/22/5, ANZ.

17 Templeton, *op cit.,* pp. 24–5.

18 W. Latta, Report on visit overseas, June–October 1955, ED1, W2675, 2/0/22/5, part 1, ANZ.

19 Cotton to McIntosh on Committee on Atomic Energy, 5 Mar. 1956, EA1, W2619, 121/2/1, ANZ.

20 Report and recommendations of the Permanent Heads Committee, n.d., ED1, W2675, 2/0/22/5, part 1, ANZ.

21 *Ibid.*

22 External Affairs paper on Nuclear Research, 15 Aug. 1956, EA1, W2619, 121/2/1, part 1, ANZ.

23 R. L. Hutchens to Minister in Charge of DSIR, 12 Mar. 1957, ED1, W2675, 2/0/22/5, ANZ.

24 *Otago Daily Times,* 9 Jan. 1958, p. 4.

25 IAEA, *History of IAEA Technical Cooperation,* http://tc.iaea.org/tcweb/abouttc/history/default.asp, downloaded 19 Feb. 2012.

26 Ian L. Baumgart, 'Hamilton, William Maxwell 1909–1992', *Dictionary of New Zealand Biography,* www.dnzb.govt.nz, downloaded 8 Apr. 2010.

27 W. M. Hamilton, report on Second United Nations International Conference on the Peaceful Uses of Atomic Energy, 10 Feb. 1959, ED1, W2673, 2/0/22/5, part 2, ANZ.

28 *The Evening Post*, 11 Nov. 1958, ED1, W2673, 2/0/22/5, part 2, ANZ.
29 J. V. Scott, Report on the Second United Nations Conference on the Peaceful Uses of Atomic Energy, ED1, W2673, 2/0/22/5, part 2, ANZ.
30 *The New Zealand Herald*, 5 Jun. 1958, p. 12.
31 *Ibid.*
32 P. N. Holloway, 'Address by Minister'. In *Symposium on Nuclear Science*, DSIR Information Series No. 23, 25 Feb. 1959.
33 Newspaper clipping, not dated or identified, but probably *The Press*, on page with clippings from April and May 1957, MB1281, Box 1, 1/1, *University of Canterbury Archives*.
34 A. D. Eppstein to USAEC, 4 Sep. 1961, CABI, CH91, 15/768, 1959–61, ANZ CHCH.
35 O'Leary to Yeabsley, 18 Sep. 1961, CABI, CH91, 15/768, 1959–61, ANZ CHCH.
36 *The Press*, 23 Oct. 1961, p. 10.
37 Email from Richard Duke, 21 Apr. 2010.
38 E. R. Collins to J. T. O'Leary, NZAEC, 25 Nov. 1960, SIR1, 74/8, vol. 4, ANZ.
39 *The New Zealand Herald*, 11 Oct. 1961, Section 1, p. 5.
40 *The Evening Post*, 28 May 1965, p. 7.

7 Nuclear decision

1 *Te Puke Times*, 17 Jun. 1955, ED1, W2673, 2/0/65/3, ANZ.
2 External Affairs briefing paper on Peaceful Uses of Atomic Energy for 1957 Prime Ministers' Conference, EA1, W2619, 121/2/1, part 1, ANZ.
3 Tony Hall, *Nuclear Politics: The History of Nuclear Power in Britain*, Penguin, Harmondsworth, 1986, p. 32.
4 W. M. Hamilton, Atomic Energy (report on 1955 overseas visit), n.d. but *c.* 1955, ED1, W2673, 2/0/22/5, ANZ.
5 *The New York Times*, 7 Aug. 1955, quoted in Canadian Nuclear Society website www.cns-snc.ca/media/toocheap/toocheap.html, downloaded 24 Nov. 2009.
6 *The Dominion*, 13 Jan. 1955, ED1, W2673, 2/0/22/5, pt 1. ANZ.
7 J. A. McWilliams, 'The Economics of Nuclear Power in New Zealand', 25 Sep. 1957, ED1, W2673, 2/0/22/5, part 2, ANZ.
8 M. G. Latta, Report on Visit Overseas June–October 1955, 17 Nov. 1955, ED1, W2673, 2/0/22/5, ANZ.

9 *Auckland Star*, 26 Apr. 1956, p. 1.
10 *The Rotorua Post*, 26 Apr. 1956, p. 6.
11 *Auckland Star*, 24 Apr. 1956, p. 3.
12 *The Dominion*, 18 Feb. 1956, EA1, W2619, 121/2/1, ANZ.
13 Darcy Walker to J. V. Wilson, Department of External Affairs, 3 Jan. 1956, EA1, W2619, 121/2/1, part 1, ANZ.
14 *Auckland Star*, 27 Apr. 1956, 2 May 1956, ED1, W2673, 2/0/65/3, ANZ.
15 *Otago Daily Times*, 14 Jun. 1956, ED1, W2673, 2/0/65/3, ANZ.
16 *The People's Voice*, 9 May 1956, ED1, W2673, 2/0/65/3, ANZ.
17 *Auckland Star*, 28 Apr. 1956, p. 3.
18 *The Evening Post*, 29 May 1957, p. 10.
19 McWilliams, *op cit.*
20 Electric Power Development: Statement by the Hon. H. Watt, 20 Aug. 1958, *AJHR* 1958, vol. 2, D4a
21 *Report of the Planning Committee on Electric Power Development in New Zealand*, NZED, 1964, p. 13.
22 Henry Coleridge Hitchcock, interview by Judith Fyfe, 19 and 23 Feb. 1987, NZOHA Electricity Centenary Oral History Project, Alexander Turnbull Library OHInt-0003/04.
23 D. H. Jones, New Zealand Nuclear Group general report on attachment to UKAEA and other related assignments, Oct. 1969 (Rob Aspden collection), pp. 8–9.
24 E. B. Mackenzie, Nuclear Fuel Power Generation, App. 4 to *Report of the Planning Committee on Electric Power Development in New Zealand*, NZED, 1964, p. 27.
25 J. T. O'Leary to Chairman and Members, NZAEC, 23 Sep. 1971, AAOQ, W3872, 72/184/6, part 1, ANZ.
26 New Zealand Oceanographic Institute to NZED, 3 Oct. 1966, ABLP W4215, 3/2, part 1, ANZ.
27 *The Dominion*, 2 Nov. 1965, p. 2.
28 *The Evening Post*, 15 May 1969, AATJ 6090, W4897/102, 5/22, part 5, ANZ.
29 *Report of the Planning Committee on Electric Power Development in New Zealand*, NZED, 1971, p. 26.
30 *Report of the Planning Committee on Electric Power Development in New Zealand*, NZED, 1974, p. 3.
31 *Ibid.*, p. 10.
32 *Ibid.*, p. 16.
33 Boyce Richardson, 'The Nuclear Decision: A Debate About New Zealand's Future: Part 1', *New Zealand Listener*, 22 Nov. 1975, pp. 14–16, at p. 14.

34 Geoff Chapple, 'Nuclear Power Slows Down', *New Zealand Listener*, 24 Feb. 1979, pp. 22–3, at p. 22.
35 E. W. Titterton and F. P. Robotham, *Uranium: Energy Source of the Future?*, Abacus, Melbourne, 1979, p. 191.
36 *The Evening Post*, 27 Jun. 1974, p. 3.
37 J. G. Gregory, 'Run Out, Fallout, All Fall Down', *New Zealand Science Review*, 31(4), 1974, p. 75.
38 Elsie Locke, *op cit.*, p. 308.
39 Richardson, *op cit.*, p. 14.
40 Hall, *op cit.*, pp. 144–5.
41 *Report of the Planning Committee on Electric Power Development in New Zealand*, NZED, 1975, p. 12.
42 Royal Commission of Inquiry into Nuclear Power Generation in New Zealand, *Nuclear Power Generation in New Zealand: Report of the Royal Commission of Inquiry*, Government Printer, Wellington, 1978, p. 20.
43 *The Dominion*, 17 Aug. 1976, p. 6.
44 Royal Commission of Inquiry into Nuclear Power Generation in New Zealand, *op cit.*, p. 36.
45 P. W. Burbidge, submission 25, Royal Commission on Nuclear Power Generation in New Zealand, Submissions, vol. 3, Government Printer, Wellington, 1978.
46 G. J. Williams, submission 31, Royal Commission on Nuclear Power Generation in New Zealand, Submissions, vol. 4, Government Printer, Wellington, 1978.
47 N. Whitehead, submission 28, *ibid*.
48 Geoff Chapple, 'Energy Which Never Ceases', *New Zealand Listener*, 28 May 1977, p. 24.
49 *Report to the New Zealand Government of the Fact Finding Group on Nuclear Power*, Government Printer, Wellington, 1977, p. 4.
50 *Ibid.*, p. 348.
51 Royal Commission of Inquiry into Nuclear Power Generation in New Zealand, *op cit.*, pp. 33, 34, 35, 45.
52 *Ibid.*, p. 45.
53 *Ibid.*, p. 41.

8 A new national identity

1 *New Zealand Parliamentary Debates*, 456 (1984), p. 256.
2 Message from Prime Minister David Lange, in *Nuclear Free New Zealand*, New Zealand Government, Wellington, 1986, p. 3.
3 *New Zealand External Affairs Review*, 14(10), 1964, p. 27.
4 L. P. Gregory, *Fallout From Nuclear Weapons Tests Conducted by France in the South Pacific from June to August 1971*, National Radiation Laboratory, Christchurch, 1972, pp. 4–5.
5 See, for example, K. M. Matthews, *Radioactive Fallout in the South Pacific: A History. Part 2: Radioactivity Measurements in the Pacific Islands*, National Radiation Laboratory, Christchurch, 1992, p. 70.
6 *New Zealand Foreign Affairs Review*, 22(6), 1972, pp. 106–7.
7 *The Dominion*, 29 Jun. 1972, p. 1.
8 *New Zealand Parliamentary Debates*, 379 (1972), p. 1119.
9 *The Evening Post*, 28 Jun. 1973, p. 1.
10 *The Evening Post*, 9 Jul. 1973, p. 1.
11 Jim McCahon, 'Voyage to Mururoa — HMNZS *Otago* and *Canterbury*', unpublished diary, 1973, p. 32.
12 Rebecca Priestley, 'Seismic Stations Play Role in Policing Nuclear Bangs', *The Evening Post*, 22 Jun. 1999, p. 5.
13 Kevin Clements, *Back From the Brink: The Creation of a Nuclear-Free New Zealand*, Allen & Unwin/Port Nicholson Press, Wellington, 1988, p. 86.
14 Barry Gustafson, *His Way: A Biography of Robert Muldoon*, Auckland University Press, Auckland, 2000, p. 230.
15 *The Dominion*, 1 Sep. 1976, p. 2.
16 Tom Newnham, *Peace Squadron: The Sharp End of Nuclear Protest in New Zealand*, Graphic Publications, Auckland, 1986, p. 12.
17 *The New Zealand Herald*, 20 Jan. 1979, p. 1.
18 *The Evening Post*, 19 Jan. 1976, p. 1.
19 *The New Zealand Herald*, 24 Mar. 1984, Section 1, p. 12
20 *Auckland Star*, 23 Mar. 1984, p. 1.
21 H. R. Atkinson *et al.*, *Report of a New Zealand, Australian and Papua New Guinea Scientific Mission to Mururoa Atoll, October–November 1983*, New Zealand Ministry of Foreign Affairs, Wellington, 1984, pp. 10–11.
22 Lange, *op cit.*, p. 31.
23 *New Zealand Parliamentary Debates*, 456 (1984), pp. 255–6.
24 Secretary of Foreign Affairs, Merwyn Norrish, in a speech to Devonport

Rotary Club on 25 Feb. 1985, *New Zealand Foreign Affairs Review*, 35(1), 1985, pp. 26–31, at p. 30.

25 Stuart McMillan, *Neither Confirm Nor Deny: The Nuclear Ships Dispute Between New Zealand and the United States*, Allen & Unwin/Port Nicholson Press, Wellington, 1987, p. 92.

26 David Lange at the Oxford Union debate, 1985, www.nzhistory.net.nz/media/sound/oxford-union-debate, downloaded 11 May 2010.

27 Michael King, *Death of the* Rainbow Warrior, Penguin, Auckland, 1986, in Author's Note.

28 *New Zealand Foreign Affairs Review*, 35(4), 1985, p. 9.

29 Clements, *op cit.*, p. 143.

30 *New Zealand Foreign Affairs Review*, 35(4), 1985, p. 4.

31 *New Zealand Foreign Affairs Review*, 36(2), 1986, p. 7.

32 Andrew McEwan, *Nuclear New Zealand: Sorting Fact From Fiction*, Hazard Press, Christchurch, 2004, p. 86.

33 Lange, *op cit.*, p. 161.

34 Eric Hobsbawm, *The Age of Extremes: A History of the World, 1914–1991*, Pantheon, New York, 1994, p. 230.

35 Robert Mann, 'Environment', in George Bryant *et al.*, *New Zealand 2001*, Cassell, Auckland, 1981, p. 52.

36 G. F. Preddey *et al.*, *Future Contingencies 4. Nuclear Disaster: A Report to the Commission for the Future by a Study Group on Nuclear Disaster*, Commission for the Future, Wellington, 1982, p. 25.

37 *Ibid.*, pp. 4, 169.

38 George Preddey, *Nuclear Disaster: A New Way of Thinking Down Under*, Asia Pacific Books: Futurewatch, Wellington, 1985, p. 5.

39 Sue McTagget, 'Defenceless', *New Zealand Listener*, 3 Dec. 1983, p. 18.

40 Preddey, *Nuclear Disaster, op cit.*, pp. 148–9.

41 P. Crutzen and J. Birks, 'The Atmosphere After a Nuclear War: Twilight At Noon', *Ambio*, 11, 1982, pp. 114–35, at p. 115.

42 R. P. Turco, O. B. Toon, T. P. Ackerman, J. B. Pollack and Carl Sagan, 'Nuclear Winter: Global Consequences of Multiple Nuclear Explosions', *Science*, 222(4630), 23 Dec. 1983, pp. 1283–92.

43 A. Barrie Pittock, *Beyond Darkness: Nuclear Winter in Australia and New Zealand*, Sun Books, Melbourne, 1987, pp. 111, 128.

44 Brian Hildreth, *A Nuclear Survival Manual for New Zealanders*, Reed Methuen, Auckland, 1986, p. 96.

45 Wren Green, Tony Cairns and Judith Wright, *New Zealand After Nuclear War*, New Zealand Planning Council, Wellington, 1987, p. 146.

Conclusion

1 *New Zealand Parliamentary Debates*, 379 (1972), p. 1001.

2 Jim McCahon, personal notes on weekend newspapers on radioactivity, May 1988, Jim McCahon's collection.

3 *The New Zealand Herald*, 11 Jul. 1962, p. 6.

SELECT BIBLIOGRAPHY

Author's note: This book is based on a PhD thesis completed as part of the University of Canterbury's History and Philosophy of Science programme. Readers wanting a view of this story with more dates, more names, more figures, full references, and a complete bibliography should refer to Rebecca Priestley, 'Nuclear New Zealand: New Zealand's Nuclear History to 1987', PhD thesis, University of Canterbury, August 2010.

ABBREVIATIONS

AJHR	Appendices to the Journals of the House of Representatives
NZJST	New Zealand Journal of Science and Technology
NZMJ	New Zealand Medical Journal
NZPD	New Zealand Parliamentary Debates
TPNZI	Transactions and Proceedings of the New Zealand Institute

PRIMARY SOURCES

UNPUBLISHED OFFICIAL RECORDS

This book draws extensively on unpublished archives, held by Archives New Zealand's Wellington and Christchurch offices, from the following government departments and organisations: Department of Scientific and Industrial Research; Ministry of Economic Development; Ministry of Foreign Affairs and Trade; Institute of Geological and Nuclear Sciences Limited; Department of External Affairs; Electricity Department; Department of Health; National Radiation Laboratory; Mines Inspection Office; The Treasury; Capital Coast Health Limited; Registrar of Companies; State Services Commission; Tourist and Publicity Department; Ministry of Works and Development. Additional archives were sourced from the University of Canterbury's records from the School of Engineering.

PUBLISHED OFFICIAL RECORDS

Published official records consulted for this book include the Social Security Act 1938, Atomic Energy Act 1945, Atomic Energy Amendment Act 1957, Electrical Wiring (X-ray) Regulations 1944, Radioactive Substances Act 1949, Radiation Protection Act 1965, Radiation Protection Regulations 1973, Transport of Radioactive Materials Regulations 1973, New Zealand Nuclear Free Zone, Disarmament and Arms Control Act 1987. Additional sources include: *New Zealand Appendices to the Journals of the House of Representatives*; *New Zealand Official Yearbook*; and *New Zealand Parliamentary Debates*.

SPECIAL PRINTED REPORTS

Academy Council of the Royal Society of New Zealand, *Radiation and the New Zealand Community: A Scientific Overview*, The Royal Society of New Zealand Bulletin 34, Wellington, 1998.

Atkinson, H. R. *et al.*, *Report of a New Zealand, Australian and Papua New Guinea Scientific Mission to Mururoa Atoll, October–November 1983*, New Zealand Ministry of Foreign Affairs, Wellington, 1984.

Crawford, John, *The Involvement of the Royal New Zealand Navy in the British Nuclear Testing Programmes of 1957 and 1958*, New Zealand Defence Force, Wellington, 1989.

Defence Committee of Enquiry, *Defence and Security: What New Zealanders Want*, Report of the Defence Committee of Enquiry, Government Printer, Wellington, 1986.

Department of Scientific and Industrial Research, *Symposium on Nuclear Science*, DSIR Information Series No. 23, Wellington, 25 February 1959.

Fact Finding Group on Nuclear Power, *Report to the New Zealand Government of the Fact Finding Group on Nuclear Power*, Government Printer, Wellington, 1977.

Grange, Les, *Prospecting for Radioactive Minerals in New Zealand*, DSIR Information Series No. 8, Government Printer, Wellington, 1954.

Gregory, L. P., *Fallout From Nuclear Weapons Tests Conducted by France in the South Pacific from June to August 1971*, National Radiation Laboratory, Christchurch, 1972.

Matthews, K. M., *Radioactive Fallout in the South Pacific: A History. Part 2: Radioactivity Measurements in the Pacific Islands*, National Radiation Laboratory, Christchurch, 1992.

Matthews, Murray, *Radioactive Fallout in the South Pacific: A History. Part 1: Deposition in New Zealand*, National Radiation Laboratory, Christchurch, 1989.

——*Radioactive Fallout in the South Pacific: A History. Part 3: Strontium-90 and Caesium-137 Deposition in New Zealand and Resulting Contamination of Milk*, National Radiation Laboratory, Christchurch, 1993.

Ministry of Civil Defence, *Public Safety Plan, Port of Wellington, Nuclear Powered Shipping*, August 1976.

Ministry of Foreign Affairs, *French Nuclear Testing in the Pacific: International Court of Justice Nuclear Tests Case, New Zealand v. France*, Ministry of Foreign Affairs publication no 446, Wellington, 1973.

—— *New Zealand at the International Court of Justice: French Nuclear Testing in the Pacific*, New Zealand Ministry of Foreign Affairs, Wellington, 1996.

National Research Bureau, *Public Opinion Poll on Defence and Security: What New Zealanders Want*, Annex to the Report of the Defence Committee of Enquiry, Wellington, 1986.

New Zealand Atomic Energy Committee, *New Zealand Code for Nuclear Powered Shipping (AEC 500)*, New Zealand Atomic Energy Committee, Wellington, 1976.

Nuclear Free New Zealand, New Zealand Government, Wellington, 1986.

Planning Committee on Electric Power Development in New Zealand, *Report of the Planning Committee on Electric Power Development in New Zealand*, New Zealand Electricity Department, Wellington, 1964–77.

Preddey, G. F. *et al.*, *Future Contingencies 4. Nuclear Disaster: A Report to the Commission for the Future by a Study Group on Nuclear Disaster*, Commission for the Future, Wellington, 1982.

Robertson, M. K., *Radioactive Waste Disposal — Policies and Practices in New Zealand*, National Radiation Laboratory Report 1996/2, Christchurch, 1996.

Roth, George, *Radiation Hazards: A Survey*, Department of Health, Wellington, 1952.

The Royal Commission into British Nuclear Tests in Australia, *The Report of the Royal Commission into British Nuclear Tests in Australia*, vol. 1, Australian Government Publishing Service, Canberra, 1985.

Royal Commission of Inquiry into Nuclear Power Generation in New Zealand, *Nuclear Power Generation in New Zealand: Report of the Royal Commission of Inquiry*, Government Printer, Wellington, 1978.

Royal Commission on Nuclear Power Generation in New Zealand, *Royal Commission on Nuclear Power Generation in New Zealand: Submissions*, vols 1–4, Government Printer, Wellington, 1978.

Special Committee on Nuclear Propulsion, *The Safety of Nuclear Powered Ships*, Department of the Prime Minister and Cabinet, Wellington, 1992.

Ward, Graham (ed.), *New Zealand Met Service Quarterly Review*, 141, New Zealand Meteorological Service, Wellington, 1984.

UNOFFICIAL RECORDS

From the Alexander Turnbull Library, the following were used: C. S. Jacobsen, Reminiscences, FMS-Papers-4253; Hutt City Council papers, MS-Papers-1293-119/03, Misc records; Sir Ernest Marsden (1889–1970) Papers, MS-Papers-1342.

This book also draws on the author's personal correspondence, either by email or face-to-face interview, from Ruth Brassington, John Campbell, Nazla Carmine (MFAT), Philip Catton, Paul Cotton, Katy Crowley (MFAT), Richard Duke, Richard Hill, Lloyd Jones, Jim McCahon and Robin Williams. It also draws on the following personal accounts:

Hitchcock, Henry Coleridge, Interview by Judith Fyfe, 19 and 23 February 1987, NZOHA
     Electricity Centenary Oral History Project, Alexander Turnbull Library OHInt-0003/04.
Jones, D. H., 'New Zealand Nuclear Group general report on attachment to UKAEA and other
     related assignments', October 1969 (Rob Aspden collection).
McCahon, Jim, Personal recollections recorded by McCahon over 14–16 August 1998 as notes
     for talk to Kaikoura Probus Club.
—— Undated press clippings and personal recollections, October 2002.
—— 'Voyage to Mururoa — HMNZSS *Otago* and *Canterbury*', unpublished diary, 1973.
Williams, Robin, Personal archives, including letter to Clement Attlee, signed by Williams and
     others, 19 September 1945.
—— 'Reflections on My Involvement in the Manhattan Project', seminar at Victoria University
     of Wellington, 10 August 2001.

NEWSPAPERS AND JOURNALS

Quotes from newspaper or journal articles are fully referenced in the endnotes. The following newspapers and journals were consulted in writing this book: *Auckland Star*, 1908, 1956, 1959, 1984; *Christchurch Star Sun*, 1956; *Christchurch Star*, 1925, 1975; *The Christchurch Sun*, 1924; *Daily Express* (Sydney), 1961; *The Daily News*, 1955; *Daily Telegraph*, 1955; *The Dominion*, 1923, 1925, 1930, 1952, 1955, 1956, 1957, 1961, 1963, 1965, 1962, 1965, 1972, 1976, 1984, 1985, 2001; *Dunedin Evening Star*, 1966; *The Evening Post*, 1932, 1939, 1955, 1956, 1957, 1958, 1961, 1962, 1965, 1973, 1974, 1976, 1977, 1985, 1999; *Gisborne Herald*, 1958; *Greymouth Evening Star*, 1972; *Here & Now*, 1955, 1957; *Manawatu Daily Times*, 1955; *The New Zealand Herald*, 1914, 1956, 1958, 1961, 1962, 1976, 1978, 1979, 1984, 1985; *New Zealand Mail*, 1896; *News*, 1999; *Otago Daily Times*, 1914, 1956, 1958; *People's Voice*, 1956; *Post*, 1974; *Press* 1898, 1900, 1902, 1905, 1914, 1924, 1953, 1955, 1956, 1957, 1961, 1997, 1954; *Rotorua Post*, 1956; *Southland News*, 1955; *Sunday Star*, 1988; *Taumarunui Press*, 1958; *Te Puke Times*, 1955; *Timaru Herald*, 1931, 1956; *Journal of the Hospital Boards Association of New Zealand*, September 1933; *New Zealand External Affairs Review*, 1954, 1955, 1956, 1963, 1964, 1968, 1990; *New Zealand Foreign Affairs Review*, 1972, 1975, 1985, 1986, 1987, 1988; *New Zealand Medical Journal*, 1896; *Student Engineer*, 1961–69; *Transactions and Proceedings of the New Zealand Institute*, 1890–1930; University of Canterbury, *School of Engineering Prospectuses*, 1963–70.

BOOKS

Boorse, Henry A. and Motz, Lloyd (eds), *The World of the Atom*, Basic Books, New York, 1966.
Marsden Editorial Committee, *Sir Ernest Marsden 80th Birthday Book*, A. H. & A. W. Reed,
     Wellington, 1969.
Needham, Joseph and Pagel, Walter, *Background to Modern Science*, Cambridge University
     Press, Cambridge, 1938.
Rutherford, Ernest, *The Collected Papers of Lord Rutherford of Nelson, vol. 1*, George Allen &
     Unwin, London, 1962.
Thomson, J. J., *Recollections and Reflections*, G. Bell and Sons, London, 1936.
Wells, H. G., *The World Set Free: A Story of Mankind*, Macmillan, London, 1914.

ARTICLES AND PAPERS

'Nuclear Testing Potential Danger', *Salient*, 29 June 1965, p. 14.

Anderson, C. C., 'Radon Ointment in Superficial Radiation Injuries', *NZMJ*, 58(323), 1959, pp. 69–71.

——'Results of X-ray and Radium Therapy at the Dunedin Hospital', *NZMJ*, 28(146), 1929, pp. 200–11.

——'The X-ray Examination of the Kidney Pelvis', *NZMJ*, 28(145), 1929, pp. 149–56.

Beck A. C., Reed J. J. and Willett R. W., 'Uranium Mineralization in the Hawks Crag Breccia of the Lower Buller Gorge Region, South Island, New Zealand', *New Zealand Journal of Geology and Geophysics*, 1(3), 1958, pp. 432–50.

Becquerel, Henri, 'On Radiations Emitted with Phosphorescence', and 'On the Invisible Radiations Emitted by Phosphorescent Bodies', *Comptes Rendus de l'Académie des Sciences*, 122 (1896), reproduced in this translation in Henry A. Boorse and Lloyd Motz (eds), *The World of the Atom*, Basic Books, New York, 1966, pp. 404–6.

Bohr, Niels, 'On the Constitution of Atoms and Molecules', *Philosophical Magazine*, 26, 1913, pp. 1–19, reproduced in Henry A. Boorse and Lloyd Motz (eds), *The World of the Atom*, Basic Books, New York, 1966, pp. 751–65.

Cameron, P. D., 'Demonstration of Radium Emanation Plant, Wellington', *NZMJ*, 28(146), 1929, pp. 240–1.

Campbell, John L., 'Thirty-six Years in Radiography', *Health and Service*, 8(1), 1953, pp. 49–52.

Cockayne, L., 'Presidential Address', *TPNZI*, 51, 1919, pp. 485–95.

Crosthwait, L. B., 'A Measurement of Atmospheric Radioactivity at Wellington', *NZJST*, 37B, 1955, pp. 382–4.

Crutzen, P. and Birks, J., 'The Atmosphere After a Nuclear War: Twilight At Noon', *Ambio*, 11, 1982, pp. 114–35.

Curie, Pierre and Curie, Marie, 'On a New Radioactive Substance Contained in Pitchblende', *Comptes Rendus de l'Académie des Sciences*, 127, 1898, pp. 175–8, reproduced in this translation in Henry A. Boorse and Lloyd Motz (eds), *The World of the Atom*, Basic Books, New York, 1966, pp. 432–4.

Curie, Pierre, Curie, Marie and Bemont, G., 'On a New Substance Strongly Radioactive, Contained in Pitchblende', *Comptes Rendus de l'Académie des Sciences*, 127, 1898, pp. 1215–17, reproduced in this translation in Henry A. Boorse and Lloyd Motz (eds), *The World of the Atom*, Basic Books, New York, 1966, pp. 434–6.

de Lautour, Harry A., 'On the Localisation of Foreign Bodies by Means of the X-rays', *NZMJ*, 1(2), 1900, pp. 67–74.

Editorial, 'A Bomb Falls', *New Zealand Listener*, 12 July 1946, p. 5.

——'Horror With Some Hope', *New Zealand Listener*, 17 August 1945, p. 5.

—— 'The Cloud That Grew,' *New Zealand Listener*, 11 April 1958, p. 10.

Evans, W. P., 'Refraction and Reflexion of X-rays', *TPNZI*, 29, 1896, pp. 573–4.

Farr, C. C. and Florance, D. C. H., 'On the Radio-activity of the Artesian Water System of Christchurch, New Zealand, and the Evidence of its Effect on Fish-life', *TPNZI*, 42, 1909, pp. 185–90.

Fieldes, M., 'Radioactivity of Niue Island Soils', *New Zealand Soil News*, 3, 1959, pp. 116–20, MS-Papers-1342-270, Alexander Turnbull Library.

Fowler, K. T., 'The "New Golden Hind" Looks for Uranium', *The Public Service Journal*, May 1946, pp. 196–9.

Grigg, F. J. T. and Rogers, M. N., 'Radioactivity and Chemical Composition of Some New Zealand Thermal Waters', *NZJST*, 11(4), 1929, pp. 216–8.

Hines, Charles F., 'The Use and Care of Radium in Hospital Practice', *The New Zealand Nursing Journal*, 34(12), 1941, pp. 400–4.

Howell, J. H., 'On the Radio-activity of Certain Soils', *TPNZI*, 39, 1906, pp. 223–6.

Hubbard, Anthony, 'MacDonell Explains His Vote', *The Dominion*, 14 June 1984, p. 1.

—— 'Muldoon Opposes Nuclear Bill', *The Dominion*, 13 June 1984, p. 2.

Laby, T. H. and Burbidge, P., 'The Nature of Gamma Rays', *TPNZI*, 44, 1911, pp. 30–1.

—— 'The Observation by Means of a String Electrometer of Fluctuations in the Ionization Produced by g Rays', *Proceedings of the Royal Society of London*, A86, 1912, pp. 333–48.

Lagan, Bernard, 'Kirk Advised Government of Return', *The Dominion*, 14 June 1984, p. 1.

Marsden, E., 'Radioactivity of Soils, Plants and Bones', *Nature*, 187(4733), 1960, pp. 192–5.

Marsden, E. and Watson-Munro, C., 'Radioactivity of New Zealand Soils and Rocks', *NZJST*, 26B(3), 1944, pp. 99–114.

Marsden, Ernest, 'Some Aspects of the Relationship of Radioactivity to Lung Cancer', *NZMJ*, 64(395), 1965, pp. 367–76.

Milligan, R. R. D. and Rogers, N. M., 'Radium Emanation and Goitre', *TPNZI*, 59, 1928, pp. 389–94.

Nicholson, D. S., 'Wartime Search for Uranium', *NZJST*, 36B, 1955, pp. 375–96.

Rafter, T. A. and Fergusson, G. J., 'The Atom Bomb Effect: Recent Increase in the 14C Content of the Atmosphere, Biosphere, and Surface Waters of the Oceans', *NZJST*, 38B(8), 1957, pp. 872–83.

Rogers, M. N., 'An Examination of the Radon and Iodine-content of Certain Christchurch Artesian and Other Waters, with Respect to the Incidence of Goitre', *TPNZI*, 57, 1927, pp. 893–9.

Roth, G. E., 'Radiation Protection: Historical Aspects and Radiographer Responsibilities', *Shadows*, 16(4), 1973, pp. 6–14.

—— 'Radiation Hazards in Perspective', *New Zealand Science Review*, 23(1), 1965, pp. 8–13.

—— 'Radiation — 1962', Health, 14(4), 1962, pp. 4–5, 10.

—— 'The Physical Services to Radiology in New Zealand', *NZMJ*, 45(248), 1946, pp. 384–8.

—— 'Radon Micro Determination by the Curtiss-Davis α-particle Counting Method', *New Zealand Journal of Science and Technology*, 27B(2), 1945, pp. 147–53.

Rutherford, E., 'Magnetic Viscosity', *TPNZI*, 28, 1895, pp. 182–204.

—— 'Magnetization of Iron by High-Frequency Discharges', *TPNZI*, 27, 1894, pp. 481–513.

—— 'The Development of the Theory of Atomic Structure', in Joseph Needham and Walter Pagel, *Background to Modern Science*, Cambridge University Press, Cambridge, 1938, pp. 61–74.

—— 'The History of Radioactivity', in Joseph Needham and Walter Pagel, *Background to Modern Science*, Cambridge University Press, Cambridge, 1938, pp. 49–60.

—— 'The Scattering of Alpha and Beta Particles by Matter and the Structure of the Atom', *Philosophical Magazine*, Series 6(21), 1911, pp. 669–88.

Rutherford, Ernest, Oliphant, Marcus and Harteck, Paul, 'Transmutation Effects Observed with Heavy Hydrogen', *Nature*, 133, 17 March 1934, p. 413.

Ryder, N.V. and Watson-Munro, C. N., 'The Detection of Radioactive Dust from the British Nuclear Bombs of October 1953', *NZJST*, 36B(2), 1954, pp. 155–9.

Stanton, Arthur, 'Wilhelm Conrad Röntgen On a New Kind of Rays: Translation of a Paper Read Before the Wurzburg Physical and Medical Society 1895', *Nature*, 53, 1896, pp. 274–6.

Thomson, J. J., 'Cathode Rays', *Philosophical Magazine*, 44, 1897, pp. 293–311, in Henry A. Boorse and Lloyd Motz (eds), *The World of the Atom*, Basic Books, New York, 1966, pp. 416–26.

Tiller, L. W. and Cooper, E. R., 'X-ray Detection of Mouldy-core in the Delicious Apple', *NZJST*, 21(3A), 1939, pp. 168–9.

Tingey, J. M. C., 'Luminizing Army Radio Equipment', *NZJST*, 27B, 1945, pp. 138–46.

Turco, R. P., Toon, O. B., Ackerman, T. P., Pollack, J. B. and Sagan, Carl, 'Nuclear Winter: Global Consequences of Multiple Nuclear Explosions', *Science*, 222(4630), 23 December 1983, pp. 1283–92.

Villard, P., 'On the Reflection and Refraction of Cathode Rays and Deviable Rays of Radium', *Comptes Rendus de l'Académie des Sciences*, 130, 1900, pp. 1010–12, reproduced in this translation in Henry A. Boorse and Lloyd Motz (eds), *The World of the Atom*, Basic Books, New York, 1966, pp. 446–8.

Wodzicki, A., 'Geochemical Prospecting for Uranium in the Lower Buller Gorge, New Zealand', *New Zealand Journal of Geology and Geophysics*, 2, 1959, pp. 602–12.

—— 'Radioactive Boulders in Hawks Crag Breccia', *New Zealand Journal of Geology and Geophysics*, 2, 1959, pp. 385–93.

X-Ray and Radium Protection Committee, 'X-Ray and Radium Protection Committee Preliminary Report', *New Zealand Journal of Health and Hospitals*, 4(9), 1921, pp. 249–51.

—— 'X-Ray and Radium Protection Committee Recommendations, Third Revised Report, Part VII', *Journal of the Hospital Boards Association of New Zealand*, 1933, pp. 24–7.

WEBSITES

Nobel Prize organisation web pages at http://nobelprize.org/ downloaded May–October 2005.
Eisenhower, Dwight D. President of the United States of America, address to the 70th Plenary Meeting of the United Nations General Assembly, www.iaea.org/About/history_speech.html, downloaded 15 September 2009.
USS *Halibut* Nuclear Submarine, www.nzhistory.net.nz/media/video/uss-halibut-nuclear-submarine, downloaded 20 April 2010.
CNN transcript of Bush's 2002 State of the Union address at http://archives.cnn.com/2002/ALLPOLITICS/01/29/bush.speech.txt/.

UNPUBLISHED

Crawford, John, 'New Zealand Observers and Indoctrinees at Nuclear Weapon Tests: 1956–1958', New Zealand Defence Force, Wellington, 2001.
Klaric, R., 'Uranium Exploration of Buller Gorge, Pororari River and Fox River Mouth Areas, New Zealand', CRA Exploration Co. Ltd, Ministry of Economic Development New Zealand, unpublished mineral report MR1250, R, 1967.
Marsden, E., 'The Atomic Age', unpublished talk to National Dairy Association of New Zealand in 1947, MS-Papers-1342-269, ATL.
Nevill, A. de T., 'Some Observations of Atmospheric Radioactivity', PhD thesis, Victoria University College, 1952.
Roth, G. E., 'The Strong Memorial Lecture 1963', unpublished lecture, National Radiation Laboratory library archives.

SECONDARY SOURCES

BOOKS

Andrade, E. N. da C., *Rutherford and the Nature of the Atom*, Peter Smith, Gloucester, Mass., 1978.
Angus, John, *A History of the Otago Hospital Board and its Predecessors*, The Otago Hospital Board, Dunedin, 1984.
Arnold, Lorna, *A Very Special Relationship: British Atomic Weapon Trials in Australia*, HMSO, London, 1987.
—— *Britain and the H-bomb*, Palgrave Macmillan, Houndmills, 2001.
Arnold, Lorna and Smith, Mark, *British Atomic Weapons Trials in Australia*, 2nd edn, Palgrave Macmillan, Houndmills, 2004.
—— *Britain, Australia and The Bomb: The Nuclear Tests and Their Aftermath*, 2nd edn, Palgrave Macmillan, Houndmills, 2006.
Badash, Lawrence, *Scientists and the Development of Nuclear Weapons: From Fission to the Limited Test Ban Treaty 1939–1963*, Humanities Press, New Jersey, 1995.
Barrowman, Rachel, *Victoria University of Wellington, 1899–1999: A History*, Victoria University Press, Wellington, 1999.
Beaglehole, John Cawte, *Victoria University College: An Essay Towards a History*, New Zealand University Press, Wellington, 1949.
Belanger, Dian Olson, *Deep Freeze: The United States, The International Geophysical Year, and the Origins of Antarctica's Age of Science*, University Press of Colorado, Boulder, 2006.
Belich, James, *Paradise Reforged: A History of the New Zealanders From the 1880s to the Year 2000*, Allen Lane, Auckland, 2001.
Bennett, F. O., *Hospital on the Avon: The History of the Christchurch Hospital 1862–1962*, North Canterbury Hospital Board, Christchurch, 1962.
Birks, J. B. (ed.), *Rutherford at Manchester*, Heywood, London, 1962.
Bolitho, Elaine E., *Reefton School of Mines, 1886–1970: Stories of Jim Bolitho*, Friends of Waiuta in association with Reefton School of Mines and the Bolitho family, Reefton, 1999.

Bothwell, Robert, *Eldorado: Canada's National Uranium Company*, University of Toronto Press, Toronto, 1994.

—— *Nucleus: The History of Atomic Energy of Canada Limited*, University of Toronto Press, Toronto, 1998.

Brocklebank, Laurie, *Jayforce: New Zealand and the Military Occupation of Japan 1945–48*, Oxford University Press, Auckland, 1997.

Broinowski, Richard, *Fact or Fission: The Truth About Australia's Nuclear Ambitions*, Scribe, Melbourne, 2003.

Bunge, Mario and Shea, William R. (eds), *Rutherford and Physics at the Turn of the Century*, Dawson and Science History Publications, New York, 1979.

Campbell, John, *Rutherford: Scientist Supreme*, AAS Publications, Christchurch, 1999.

Carey, John (ed.), *The Faber Book of Science*, Faber, London, 1995.

Cathcart, Brian, *The Fly in the Cathedral: How a Small Group of Cambridge Scientists Won the Race to Split the Atom*, Viking, London, 2004.

Caulfield, Catherine, *Multiple Exposures: Chronicles of the Radiation Age*, Penguin, London, 1989.

Cawte, Alice, *Atomic Australia, 1944–1990*, New South Wales University Press, Sydney, 1992.

Clarke, Margaret (ed.), *For The Record: Lange and the Fourth Labour Government*, Dunmore Publishing, Palmerston North, 2005.

Clements, Kevin, *Back From the Brink: The Creation of a Nuclear-Free New Zealand*, Allen & Unwin/Port Nicholson Press, Wellington, 1988.

Cross, Roger, *Fallout: Hedley Marston and the British Bomb Tests in Australia*, Wakefield Press, Kent Town, 2001.

Danielsson, Bengt and Danielsson, Marie-Therese, *Poisoned Reign: French Nuclear Colonialism in the Pacific*, Penguin, Ringwood, Vic., 1986.

Dewes, Kate and Green, Robert, *Aotearoa/New Zealand at the World Court*, The Raven Press, Christchurch, 1999.

Dow, Derek A., *Safeguarding the Public Health: A History of the New Zealand Department of Health*, Victoria University Press, Wellington, 1995.

Duncan, T. and Stout, M., *Medical Services in New Zealand and the Pacific*, War History Branch, Department of Internal Affairs, Wellington, 1958.

—— *War Surgery and Medicine*, War History Branch, Department of Internal Affairs, Wellington, 1954.

Dyson, John, *Sink the Rainbow! An Enquiry into the 'Greenpeace Affair'*, Reed Methuen, Auckland, 1986.

Evans, Ivor B. N., *Man of Power: The Life Story of Baron Rutherford of Nelson*, Stanley Paul, London, 1939.

Eve, A. S., *Rutherford: Being the Life and Letters of the Rt Hon Lord Rutherford, O. M.*, Cambridge University Press, Cambridge, 1939.

Fairburn, Miles, *The Ideal Society and Its Enemies*, Auckland University Press, Auckland, 1989.

Feather, Norman, *Lord Rutherford*, Blackie & Son, London, 1940.

Fenwick, P. C., *North Canterbury Hospital and Charitable Aid Board*, Andrews Baty, Christchurch, 1926.

Firth, Stewart, *Nuclear Playground*, Allen & Unwin, Sydney, 1987.

Gaddis, John Lewis, *The Cold War*, Allen Lane, London, 2005.

Galbreath, Ross, *DSIR: Making Science Work for New Zealand*, Victoria University Press, Wellington, 1998.

Gardner, W. J., Beardsley, E. T. and Carter, T. E., *A History of the University of Canterbury, 1873–1973*, University of Canterbury, Christchurch, 1973.

Gowing, Margaret, *Britain and Atomic Energy 1939–1945*, Macmillan, London, 1964.

—— *Independence and Deterrence: Britain and Atomic Energy 1945–52, Volume 1, Policy Making*, Macmillan, London, 1974.

—— *Independence and Deterrence: Britain and Atomic Energy 1945–52, Volume 2, Policy Execution*, Macmillan, London, 1974.

Green, Wren, Cairns, Tony and Wright, Judith, *New Zealand After Nuclear War*, New Zealand Planning Council, Wellington, 1987.

Gribbin, John, *Science: A History 1543–2001*, Allen Lane, London, 2002.

Gustafson, Barry, *His Way: A Biography of Robert Muldoon*, Auckland University Press, Auckland, 2000.

Hall, Tony, *Nuclear Politics: The History of Nuclear Power in Britain*, Penguin, Harmondsworth, 1986.

Harremoes, Paul (ed.), *The Precautionary Principle in the 20th Century: Late Lessons from Early Warnings*, Earthscan Publications, London, Sterling VA, 2000.

Hayman, Maurice, *These Useless Wings: Operation Grapple*, M. Hayman, Riverton, 1997.

Hellemans, Alexander and Bunch, Bryan, *The Timetables of Science: A Chronology of the Most Important People and Events in the History of Science*, Simon and Schuster, New York, 1988.

Hercus, Sir Charles and Bell, Sir Gordon, *The Otago Medical School under the First Three Deans*, E & S Livingstone, Edinburgh and London, 1964.

Hewlett, Richard G. and Holl, Jack M., *Atoms for Peace and War, 1953–1961: Eisenhower and the Atomic Energy Commission*, University of California Press, Berkeley, 1989.

Hildreth, Brian, *A Nuclear Survival Manual for New Zealanders*, Reed Methuen, Auckland, 1986.

Hobsbawm, Eric, *The Age of Extremes: A History of the World, 1914–1991*, Pantheon, New York, 1994.

Jamieson, H. D. (ed.), *The Development of Medical Physics and Biomedical Engineering in New Zealand Hospitals 1945–1995*, self-published, Dannevirke, 1995.

Kellaway, Jo and Maryan, Mike, *A Century of Care: Palmerston North Hospital 1893–1993*, Focus Books, Double Bay, 1993.

Keyler, William R., *The Twentieth Century World: An International History*, Oxford University Press, New York, Oxford, 1996.

King, Michael, *Death of the* Rainbow Warrior, Penguin, Auckland, 1986.

—— *The Penguin History of New Zealand*, Penguin, Auckland, 2003.

Lange, David, *Nuclear Free — the New Zealand Way*, Penguin Books, Auckland, 1990.

Locke, Elsie, *Peace People: A History of Peace Activities in New Zealand*, Hazard Press, Christchurch, 1992.

Maclean, F. S., *Challenge for Health: A History of Public Health in New Zealand*, Government Printer, Wellington, 1964.

McEwan, A. C., *Radiation Protection and Dosimetry in New Zealand: A History of the National Radiation Laboratory*, New Zealand Department of Health, Wellington, 1983.

McEwan, Andrew, *Nuclear New Zealand: Sorting Fact From Fiction*, Hazard Press, Christchurch, 2004.

McGibbon, Ian (ed.), *Undiplomatic Dialogue: Letters Between Carl Berendsen and Alister McIntosh, 1943–52*, Auckland University Press, Auckland, 1993.

McKinnon, Malcolm. *Independence and Foreign Policy: New Zealand in the World Since 1935*, Auckland University Press, Auckland, 1993.

McLintock, A. H. (ed.), *An Encyclopaedia of New Zealand*, Vol. 2, Government Printer, Wellington, 1966.

McMillan, Stuart, *Neither Confirm Nor Deny: The Nuclear Ships Dispute Between New Zealand and the United States*, Allen & Unwin/Port Nicholson Press, Wellington, 1987.

Mann, W. B., *The Cyclotron*, 4th edn, Methuen, London, 1953.

Martin, John E. (ed.), *People, Politics and Power Stations: Electric Power Generation in New Zealand 1880–1998*, 2nd edn, Electricity Corporation of New Zealand and Historical Branch, Department of Internal Affairs, Wellington, 1998.

May, Philip Ross, *The West Coast Gold Rushes*, Pegasus, Christchurch, 1967.

Mayell, Mark, *Nuclear Accidents*, Lucent Books, Farmington Hills, 2004.

Meduna, Veronika and Priestley, Rebecca, *Atoms, Dinosaurs & DNA: 68 Great New Zealand Scientists*, Random House, Auckland, 2008.

Medvedev, Zhores A., *The Legacy of Chernobyl*, W. W. Norton, New York, 1990.

Mein Smith, Philippa, *A Concise History of New Zealand*, Cambridge University Press, Cambridge, 2005.

—— *Maternity in Dispute: New Zealand 1920–1939*, Department of Internal Affairs, Wellington, 1986.

Miles, Julie and Shaw, Elaine, *Chronology: The French Presence in the South Pacific, 1838–1990*, Greenpeace, Auckland, 1990.

Mitchell, Alan, *84 Not Out: The Story of Sir Arthur Sims*, Hennel Locke in association with George G. Harrap & Co., London, 1962.

Morgan, Robin and Whitaker, Brian, *The Sunday Times Insight Rainbow Warrior*, Arrow Books, London, Auckland, 1986.

Morrell, W. P., *The University of Otago: A Centennial History*, University of Otago Press, Dunedin, 1969.

Muirhead, Ed., *A Man Ahead of His Times: T. H. Laby's Contribution to Australian Science*, University of Melbourne, Melbourne, 1996.

Neutze, Diana and Beardsley, Eric, *Design for a Century*, University of Canterbury Publication No. 37, Christchurch, 1987.

Newnham, Tom, *Peace Squadron: The Sharp End of Nuclear Protest in New Zealand*, Graphic Publications, Auckland, 1986.

Oliphant, Mark, *Rutherford: Recollections of the Cambridge Days*, Elsevier, Amsterdam, New York, 1972.

Palmer, Geoffrey and Palmer, Matthew, *Bridled Power: New Zealand Government Under MMP*, 3rd edn, Oxford University Press, Auckland, 1997.

Parry, Gordon, *Clipping the Claws: The Story of the First 60 years of the Otago and Southland Division of the Cancer Society of New Zealand*, Otago and Southland Division of the Cancer Society of New Zealand, Dunedin, 1989.

Peat, Neville, *Subantarctic New Zealand: A Rare Heritage*, Department of Conservation, Invercargill, 2003.

Pittock, A. Barrie, *Beyond Darkness: Nuclear Winter in Australia and New Zealand*, Sun Books, Melbourne, 1987.

Porter, Roy and Ogilvie, Marilyn (eds), *The Biographical Dictionary of Scientists, Volume 3*, 3rd edn, Oxford University Press, New York, 2000.

Powaski, Ronald E., *March to Armageddon: The United States and the Nuclear Arms Race, 1939 to the Present*, Oxford University Press, Oxford, 1987.

Preddey, George, *Nuclear Disaster: A New Way of Thinking Down Under*, Asia Pacific Books: Futurewatch, Wellington, 1985.

Priestley, Rebecca (ed.), *The Awa Book of New Zealand Science*, Awa Press, Wellington, 2008.

Quinn, Susan, *Marie Curie: A Life*, Heinemann, London, 1995.

Reynolds, Wayne, *Australia's Bid for the Atomic Bomb*, Melbourne University Press, Melbourne, 2000.

Rhodes, Richard, *The Making of the Atomic Bomb*, Touchstone, New York, 1986.

Rice, Geoffrey W. (ed.), *The Oxford History of New Zealand*, 2nd edition, Oxford University Press, Auckland, 1992.

Rockel, Ian, *Taking the Waters: Early Spas in New Zealand*, Government Print, Wellington, 1986.

Ryan, James, Sutton, Keith and Baigent, Malcolm, *Australasian Radiology: A History*, McGraw-Hill, Sydney, 1996.

Shears, Richard and Gidley, Isobelle, *The* Rainbow Warrior *Affair*, Irwin Publishing, Toronto, 1986.

Shurcliff, W. A., *Bombs at Bikini*, W. H. Wise, New York, 1947.

Sinclair, Keith, *A History of the University of Auckland 1883–1983*, Auckland University Press, Auckland, 1983.

Spencer R. Weart, *Nuclear Fear: A History of Images*, Harvard University Press, Cambridge, Mass., 1988.

Sutphen, Mary P. and Andrews, Bridie (eds), *Medicine and Colonial Identity*, Rutledge, London and New York, 2003.

Szabo, Michael, *Making Waves: The Greenpeace New Zealand Story*, Greenpeace New Zealand, Auckland, 1991.

Taylor, Nancy, *The New Zealand People at War: The Home Front, Volume 2*, Government Printer, Wellington, 1986.

Templeton, Malcolm, *Standing Upright Here: New Zealand in the Nuclear Age 1944–1990*, Victoria University Press, Wellington, 2006.

Trapeznik, Alexander and Fox, Aaron (eds), *Lenin's Legacy Down Under: New Zealand's Cold War*, University of Otago Press, Dunedin, 2004.

Trotter, Chris, *No Left Turn: The Distortion of New Zealand's History by Greed, Bigotry and Right-Wing Politics*, Random House, Auckland, 2007.

Walker, J. Samuel, *Permissible Dose: A History of Radiation Protection in the Twentieth Century*, University of California Press, Berkeley, 2000.

Wilson, D. Macdonald, *A Hundred Years of Healing: Wellington Hospital 1847–1947*, A. H. & A. W. Reed, Wellington, 1948.

Wilson, David, *Rutherford: Simple Genius*, The MIT Press, Cambridge, Mass., 1983.

ARTICLES, PAPERS AND CHAPTERS

'Atomic Energy in Harness', *The Economist*, 23 July 1955, pp. 1–17.

'New Zealand Strontium-90 Figures "A warning" ', *New Zealand Public Service Journal*, December 1959, p. 8.

'X-Day at Bikini', *New Zealand Listener*, 5 July 1946, pp. 6–8.

Anderson, C. C., 'The Development of Radiology in New Zealand', *Australasian Radiology*, 10, 1966, pp. 296–307.

Avery, Donald Howard, 'Atomic Scientific Co-operation and Rivalry Among Allies: The Anglo-Canadian Montreal Laboratory and the Manhattan Project 1943–1946', *War in History*, 2(3), 1995, pp. 274–305.

Barber, David, 'New Zealand Frigate Making Headlines in Europe as She Continues to Defy France', *The Evening Post*, 11 July 1973, p. 1.

Beavis, Cindy, 'Handling "Hot" Rubbish', *New Zealand Listener*, 22 October 1976, pp. 16–17.

Begg, A. C., 'The Father of Radiology in New Zealand', *NZMJ*, 83(543), 1975, pp. 1–5.

Binnie, Anna, 'Oliphant, the Father of Atomic Energy', *Journal and Proceedings of the Royal Society of New South Wales*, 139, 2006, pp. 11–22.

Borrie, J. F., 'A History of Training for the Diploma of the Society of Radiographers (London)', *Shadows*, 16(1), 1973, pp. 15–27.

Brady, Anne-Marie, 'The War That Never Was: New Zealand–China Relations in the Cold War Era', in Alexander Trapeznik and Aaron Fox (eds), *Lenin's Legacy Down Under: New Zealand's Cold War*, University of Otago Press, Dunedin, 2004, pp. 131–51.

Brown, Shaun, 'On Protest with Otago', *New Zealand Listener*, 20 August 1973, p. 14.

Bryder, Lynda, 'Tuberculosis in New Zealand', in A. J. Proust (ed.), *History of Tuberculosis in Australia, New Zealand and Papua New Guinea*, Brolga Press, Canberra, 1991, pp. 79–89.

Calcott, Dean, 'Strontium Powered, Antarctica's Nuclear Past is Little Known, *The Press*, 27 December 1996, p. 14.

Calder, Ritchie, 'Burying Live Atoms is as Expensive as Burying Dead Pharoahs', *WHO Special Feature*, November 1958.

Campbell, Diane J., 'A Brief History of Dental Radiography', *New Zealand Dental Journal*, 91, 1995, pp. 127–33.

Chapple, Geoff, 'Energy Which Never Ceases', *New Zealand Listener*, 28 May 1977, pp. 24–5.

—— 'Nuclear Power Slows Down', *New Zealand Listener*, 24 February 1979, pp. 22–3.

Crawford, John (ed.), *Kia Kaha: New Zealand in the Second World War*, Oxford University Press, Melbourne, 2002.

Crawford, John, ' "A Political H-Bomb": New Zealand and the British Thermonuclear Weapon Tests of 1957–58', *The Journal of Imperial and Commonwealth History*, 26(1), 1998, pp. 127–50.

Curie, Eve, 'The Colour of Radium', in John Carey (ed.), *The Faber Book of Science*, Faber, London, 1995, pp. 191–201.

Dow, Derek, 'Electricians Left to Run Early X-ray Machines', *New Zealand Doctor*, 30 September 1998, p. 35.

DSIR, *Symposium on Nuclear Science*, DSIR Information Series No. 23, Wellington, 25 February 1959.

Editorial, 'The Deadly Dust', *New Zealand Listener*, 29 June 1956, p. 4.

Farley, F. J. M., 'The Fewer the Better', *Here & Now*, 60, September 1957, pp. 11–12.

Fermi, Laura, 'In the Black Squash Court', in John Carey (ed.), *The Faber Book of Science*, Faber, London, 1995, pp. 324–34.

Gair, George, 'Muldoon and His Cabinet', in Margaret Clark (ed.), *Muldoon Revisited*, Dunmore Press, Palmerston North, 2004, pp. 41–57.

Galbreath, Ross, 'The Rutherford Connection: New Zealand Scientists and the Manhattan and Montreal Projects', *War in History*, 2(3), 1995, pp. 306–19.

Goldschmidt, Bertrand, 'A Historical Survey of Nonproliferation Policies', *International Security*, 2(1), 1977, pp. 69–87.

Gregory, J. G., 'Run Out, Fallout, All Fall Down', *New Zealand Science Review*, 31(4), 1974, p. 75.

Gregory, L. P., 'Radioactive Fallout', *Health and the Environment*, 24(4), 1972, pp. 14–15.

Henderson, John, 'The Warrior Peacenik: Setting the Record Straight on ANZUS and The Fiji Coup', in Margaret Clarke (ed.), *For The Record: Lange and the Fourth Labour Government*, Dunmore Publishing, Palmerston North, 2005, pp. 136–43.

Hensley, Gerald, 'The Bureaucracy and Advisors', in Margaret Clarke (ed.), *For The Record: Lange and the Fourth Labour Government*, Dunmore Publishing, Palmerston North, 2005, pp. 129–35.

Holland, Sidney, 'Nuclear Test Explosions', *External Affairs Review*, 7(5), 1957, pp. 17–20.

Knox, W. A., 'Protective Measures in Dental Roentgenology', *New Zealand Dental Journal*, March 1920, pp. 136–9.

Lake, Marilyn, 'Female Desires: The Meaning of World War II', *Australian Historian Studies*, 24(95), 1990, pp. 267–84.

Mann, Robert, 'Environment', in George Bryant *et al.*, *New Zealand 2001*, Cassell, Auckland, 1981, p. 52.

Mann, Robert and Owen Wilkes, 'The Story of Nukey-Poo', *The Bulletin of the Atomic Scientists*, 34(8), 1978, pp. 32–6.

McIntyre, W. David, 'From Dual Dependency to Nuclear Free', in Geoffrey W. Rice (ed.), *The Oxford History of New Zealand*, 2nd edn, Oxford University Press, Auckland, 1992, pp. 520–38.

McTagget, Sue, 'Defenceless', *New Zealand Listener*, 3 December 1983, p. 18.

Mein Smith, Philippa, 'New Zealand Milk for "Building Britons"', in Mary P. Sutphen and Bridie Andrews (eds), *Medicine and Colonial Identity*, Rutledge, London and New York, 2003, pp. 79–102.

National Radiation Laboratory, 'Peaceful Plutonium', *Radiation Protection News & Notes*, 34, April 1996, pp. 7–8.

Nicholson, D. S., 'Wartime Search for Uranium', *NZJST*, 36B, 1955, pp. 375–96.

Norrish, Merwyn, 'The Lange Government's Foreign Policy', in Margaret Clarke (ed.), *For The Record: Lange and the Fourth Labour Government*, Dunmore Publishing, Palmerston North, 2005, pp. 150–7.

O'Meagher, Matthew, 'Prospects for Enrichment: New Zealand Responses to the Peaceful Atom in the 1950s', *Journal & Proceedings of the Royal Society of New South Wales*, 139, 2006, pp. 51–62.

Pauling, Linus, 'Atomic Power and Radiation Hazards', *New Zealand Listener*, 18 December 1959, pp. 6–7.

Philips, Jock, 'New Zealand Celebrates Victory', in John Crawford (ed.), *Kia Kaha: New Zealand in the Second World War*, Oxford University Press, Melbourne, 2002, pp. 302–16.

Priestley, Rebecca, 'Ernest Marsden's Nuclear New Zealand: from Nuclear Reactors to Nuclear Disarmament', *Journal and Proceedings of the Royal Society of New South Wales*, 139, 2006, pp. 23–38.

—— 'Lord of the Atoms', *New Zealand Listener*, 15 November 2008, pp. 28–32.

—— 'Material Evidence', *New Zealand Listener*, 27 August 2005, pp. 36–7.

—— 'Seismic Stations Play Role in Policing Nuclear Bangs', *Evening Post*, 22 June 1999, p. 5.

—— 'The Search for Uranium in "Nuclear-free" New Zealand: Prospecting on the West Coast, 1940s to 1970s', *New Zealand Geographer*, 62, 2006, pp. 121–4.

Rafter, Athol, Extract from 'Problems in the Establishment of a Carbon-14 and Tritium Laboratory', paper presented at the Sixth International Conference on Radiocarbon and Tritium Dating, Washington State University, 1965, reproduced in Rebecca Priestley (ed.), *The Awa Book of New Zealand Science*, Wellington, Awa Press, 2008, pp. 250–4.

Reid, Francis, ' "The Democratic Politician Does Not Trouble Himself With Science": Class and Professionalism in the New Zealand Institute, 1867–1903', *Tuhinga*, 16, 2005, pp. 16–31.

Richardson, Boyce, 'The Nuclear Decision: A Debate About New Zealand's Future: Part 1', *New Zealand Listener*, 22 November 1975, pp. 14–16.

Robie, David, 'The Day of Two Suns', *New Zealand Listener*, 10 August 1985, pp. 18–19.

Ross, Larry, 'Nuclear Flashback' (letter to the editor), *New Zealand Listener*, 7 September 1985, p. 12.

Roth, G. E., 'The Problem of Radiation Safety', *The New Zealand Radiographer*, 1(4), 1949, pp. 3–14.

Sparks, Rodger, 'Radiocarbon Dating — New Zealand Beginnings', *New Zealand Science Review*, 61(2), 2004, pp. 39–41.

Strange, Glyn, 'Popper's A-bomb Dilemma', *Evening Post*, 14 August 1995, p. 5.

Weiner, Charles, 'Physics in the Great Depression', *Physics Today*, 23(10), 1970, pp. 31–8.

Wilkes, Owen, 'New Zealand and the Atom Bomb', in John Crawford (ed.), *Kia Kaha: New Zealand in the Second World War*, Oxford University Press, Melbourne, 2002, pp. 264–75.

Williamson, B. D. P., 'Luminous Paint', *N. Z. Horological Journal*, 17(5), 1963, pp. 17–23.

—— 'Luminous Paint: No Significant Hazards Detected', *N. Z. Horological Journal*, 14(4), 1960, pp. 6–9.

Wood, F. L. W., 'New Zealand Foreign Policy 1945–1951', in *New Zealand in World Affairs, volume 1*, Price Milburn/New Zealand Institute of International Affairs, Wellington, 1977, pp. 91–113

Yeabsley, H. J., 'No Danger in N.Z. from Atomic Fallout', *Parent and Child*, 3(5), 1960, p. 26.

WEBSITES AND ELECTRONIC MEDIA

AAS biography of Watson-Munro, www.science.org.au/academy/memoirs/watson-munro.htm, downloaded 10 December 2009.

American Institute of Physics internet entry on J. J. Thomson at www.aip.org/history/electron/jjhome.htm, downloaded 23 June 2005.

ANSTO website, www.ansto.gov.au/discovering_ansto/history_of_ansto/hifar, downloaded 20 April 2010.

Baumgart, Ian L., 'Hamilton, William Maxwell 1909–1992', *Dictionary of New Zealand Biography*, www.dnzb.govt.nz, downloaded 8 April 2010.

Brennan, M. H., 'Charles Watson-Munro, 1915–1991', Australian Academy of Science Biographical Memoirs, www.science.org.au/academy/memoirs/watson-munro.htm, downloaded 17 September 2008.

Brodie, James W., 'Howell, John Henry 1869–1944', *Dictionary of New Zealand Biography*, www.dnzb.govt.nz, downloaded 21 June 2005.

Bush, Graham W. A., 'Bloodworth, Thomas 1882–1974', *Dictionary of New Zealand Biography*, www.dnzb.govt.nz/, downloaded 11 June 2010.

Clark, Helen, Prime Minister's address on 'New Zealand and Peaceful Conflict Resolution' in Cairo, Egypt, www.beehive.govt.nz/speech/new+zealand+and+peaceful+conflict+resoluti on, downloaded 23 June 2010.

Department of Conservation website: www.doc.govt.nz/Conservation/Marine-and-Coastal/Marine-Reserves/Kermadec.asp, downloaded 15 July 2004

Dow, Derek A., 'Watt, Michael Herbert 1887–1967', *Dictionary of New Zealand Biography*, www.dnzb.govt.nz, downloaded 20 June 2010.

Dunsford, Deborah Ann, 'Seeking the Prize of Eradication: A social history of tuberculosis in New Zealand from World War Two to the 1970s', PhD thesis, University of Auckland, 2008, https://researchspace.auckland.ac.nz/handle/2292/2932, downloaded 3 January 2012.

Exchange rates over time calculated at www.measuringworth.com and at www.usinflationcalculator.com.

Fischer, David, *History of the IAEA: The First Forty Years*, Vienna, The Agency, 1997, pp. 471–4, www-pub.iaea.org/MTCD/publications/PDF/Pub1032_web.pdf, downloaded 21 October 2009.

Frame, Paul W., 'Radioactive Curative Devices and Spas', originally published in the Oak Ridger newspaper, 5 November 1989, www.orau.org/ptp/articlesstories/quackstory.htm, downloaded 1 June 2004.

Full text of the Convention on the High Seas Done at Geneva, on 29 April 1958, http://sedac.ciesin.org/entri/texts/high.seas.1958.html, downloaded 14 December 2009.

Galbreath, Ross, 'Marsden, Ernest 1889–1970', *Dictionary of New Zealand Biography*, www.dnzb.govt.nz/, downloaded 21 November 2005.

Goldschmidt, Dr Bertrand, 'Uranium's Scientific History 1789–1939', originally presented at the 14th International Symposium held by the Uranium Institute in London, September 1989, www.world-nuclear.org/ushist.htm, downloaded 27 June 2005.

Gustafson, Barry, 'Holland, Sidney George 1893–1961', *Dictionary of New Zealand Biography*, www.dnzb.govt.nz/, downloaded 8 March 2010.

Home, R. W., 'Burhop, Eric Henry Stoneley (1911–1980), *Australian Dictionary of Biography*, http://adbonline.anu.edu.au/biogs/A130339b.htm, downloaded 6 December 2008.

IAEA, *History of IAEA Technical Cooperation*, http://tc.iaea.org/tcweb/abouttc/history/default. asp, downloaded 19 February 2012.

Lange, David, at the Oxford Union debate, 1985, www.nzhistory.net.nz/media/sound/oxford-union-debate, downloaded 11 May 2010.

Marshall, Russell, 'Bledisloe, Charles Bathurst 1867–1958', *Dictionary of New Zealand Biography*, www.dnzb.govt.nz/, downloaded 12 June 2010.

McLintoch, A. H. (ed.), 'North and South Island Electrical Interconnection Including the Cook Strait Cable', An Encyclopedia of New Zealand, 1966, www.teara.govt.nz/en/1966/power-resources/10, downloaded 23 November 2009.

—— 'Recent Research', *An Encyclopaedia of New Zealand*, 1966, www.teara.govt.nz/en/1966/ nuclear-science-in-new-zealand/4, downloaded 15 March 2010.

Ministry of Civil Defence, *Civil Defence in New Zealand: A Short History,* Ministry of Civil Defence, Wellington, 1990, napier.digidocs.com/userfiles/file/cd_short_history.pdf, downloaded 10 February 2010.

MIT Inventor of the Week entry on Johannes Geiger at http://web.mit.edu/invent/iow/geiger. html, downloaded 22 August 2005.

New York Times, 7 August 1955, quoted in Canadian Nuclear Society website www.cns-snc.ca/ media/toocheap/toocheap.html, downloaded 24 November 2009.

Nuclear test yields from www.ratical.org/ratville/nukes/testChrono95-8.html, downloaded 12 December 2004.

NZPA, 'Reactor Tempers Hutt City's Nuke Fee Declaration', www.rsnz.govt.nz/archives/news_ pre_oct99/news/y_1995/m _09/d_18/a_3.php, downloaded 12 January 2000.

—— 'Uranium ore passes through NZ ports once a week, says ministry', http://www.stuff.co.nz/ dominion-post/national/3847692/Uranium-ore-passes-through-NZ-ports-once-a-week-says-ministry, downloaded 25 June 2010.

Oak Ridge Associated Universities Internet article on 'Shoe-fitting Fluoroscope (ca. 1930–1940)', www.oran.org/ptp/collection/shoefittingfluor/shoe.htm, downloaded 2 June 2004.

Parliamentary debate on New Zealand Nuclear Free Zone, Disarmament, and Arms Control (Nuclear Propulsion Reform) Amendment Bill at http://www.parliament.nz/miNZ/PB/ Debates/Debates/7/c/1/47HansD_20050727_00001577-New-Zealand-Nuclear-Free-Zone-Disarmament.htm, downloaded 23 July 2008.

Priestley, Rebecca. 'The Search for Uranium in New Zealand', *Te Ara Encyclopaedia of New Zealand*, www.teara.govt.nz/earthseaandsky/mineralresources/radioactiveminerals/en, downloaded 4 July 2007.

Richardson, Keith, *Fatal Legacy: A Nuclear Story*, Alexander Turnbull Library Sound Recordings Collection Phono CD9755.

Simon, S. L. and W. L. Robison, 'A compilation of nuclear weapons test detonation data for U.S. Pacific Ocean tests', *Journal of Health Physics*, 73(1), 1997, pp. 258–64, www.hss.energy. gov/HealthSafety/IHS/marshall/marsh/journal/rpt-22.pdf, downloaded 26 March 2009.

Thain, Iain A., *A Brief History of the Wairakei Geothermal Power Project*, http://geoheat.oit. edu/bulletin/bull19-3/art1.pdf, downloaded 27 November 2004, reproduced from Geo-Heat Bulletin, 19, 3.

Topp, Jools, The Topp Twins interview, 'Nine to Noon', Tuesday, 31 March 2009, www.radionz. co.nz/national/programmes/ninetonoon/20090331, downloaded 21 March 2010.

United Nations website, www.un.org/aboutun.milestones.htm.

UNSCEAR official website, www.unscear.org/unscear/en/about_us/history.html, downloaded 15 December 2009.

USAEC, Report on Project Gabriel, July 1954, www.hss.energy.gov/HealthSafety/IHS/marshall/ collection/data/1hp1b/5674_.pdf , downloaded 15 December 2009.

Wilson, James Oakley, 'Holland, Sir Sidney George', from A. H. McLintock (ed.), *An Encyclopaedia of New Zealand*, 1966, www.teara.govt.nz/en/1966/holland-sir-sidney-george/1, downloaded 8 March 2010.

INDEX

*Page numbers in **bold** refer to images/captions.*